Communicating Emergency Preparedness

Strategies for Creating a Disaster Resilient Public

OTHER CRC PRESS TITLES

Pandemic Influenza: Emergency Planning and Community Preparedness
Jeffrey Ryan
ISBN: 1-4200-6087-2, Catalog Number: 60872

Medical Disaster Response: A Survival Guide for Hospitals in Mass Casualty Events
Robert Bonvino and David Goldschmitt
ISBN: 1-4200-6122-4, Catalog Number: 61224

Natural Hazards Analysis: Reducing the Impact of Disasters
John Pine
ISBN: 1-4200-7038-X, Catalog Number: AU7038

Global Warming, Natural Hazards, and Emergency Management
Jane A. Bullock, George D. Haddow and Kim S. Haddow
ISBN: 1-4200-8182-9, Catalog Number: AU1829

The Human Side of Disaster
Thomas Drabek
ISBN: 1-4398-0864-3, Catalog Number: K10428

Social Vulnerability to Disasters
Brenda Phillips, et al.
ISBN: 1-4200-7856-9, Catalog Number: AU7856

Disaster Recovery
Brenda Phillips
ISBN: 1-4200-7420-2, Catalog Number: AU7420

Building an Enterprise Business Continuity Program
Patricia Okolita
ISBN: 1-4200-8864-5, Catalog Number: AU8645

Disaster Management Handbook
Jack Pinkowski
ISBN: 1-4200- 5862-2, Catalog Number: AU5862

Communicating Emergency Preparedness

Strategies for Creating a Disaster Resilient Public

Damon P. Coppola • Erin K. Maloney

CRC Press
Taylor & Francis Group
Boca Raton London New York

CRC Press is an imprint of the
Taylor & Francis Group, an **Informa** business

Auerbach Publications
Taylor & Francis Group
6000 Broken Sound Parkway NW, Suite 300
Boca Raton, FL 33487-2742

Printed in the United States of America on acid-free paper
10 9 8 7 6 5 4 3 2 1

International Standard Book Number: 978-1-4200-6510-7 (Hardback)

Library of Congress Cataloging-in-Publication Data

Coppola, Damon P.
Communicating emergency preparedness : strategies for creating a disaster resilient public / Damon Coppola, Erin K. Maloney.
p. cm.
Includes bibliographical references and index.
ISBN 978-1-4200-6510-7 (hbk. : alk. paper)
1. Emergency management. 2. Risk communication. I. Maloney, Erin K. II. Title.

HV551.2.C635 2009
363.34'7--dc22 2009014256

Visit the Taylor & Francis Web site at
http://www.taylorandfrancis.com

and the Auerbach Web site at
http://www.auerbach-publications.com

DEDICATION

The authors dedicate this book to Eileen and Paul Coppola, and Nancy and Tim Maloney, from whom countless risk lessons have come.

CONTENTS

ACKNOWLEDGMENTS

Damon Coppola would like to express his profound gratitude to his wife, Mary Gardner Coppola, for her guidance and patience; to Jane Bullock and George Haddow for their friendship and support and for continuing to give freely of their expertise and experience; to Eileen Coppola for her sage editing assistance; and to Jack Harrald, Greg Shaw, J. René van Dorp, and Ollie Davidson for their insight.

Erin Maloney would like to thank her former advisors Dr. Dana Mastro and Dr. Leslie Snyder for inspiring her to go into the field; her current doctoral advisor, Dr. Maria Lapinski, for her guidance and support; and Kyle for teaching her the value of home-roasted coffee.

The authors would also like to thank Mark Listewnik, Jessica Vakili, and Andrea Demby at Taylor & Francis for the assistance they provided in the development of this book.

INTRODUCTION

The first eight years of the 21st century were punctuated by a diverse series of mega-disasters, most notably the earthquakes in Gujarat, India, and Kashmir, Pakistan, the September 11 terrorist attacks on the United States, the Indian Ocean tsunami, Hurricane Katrina, Cyclone Nargis in Burma, and the Sichuan earthquake in China. After-action reporting following these events denoted that only a very small percentage of the affected populations had acted to reduce their vulnerability prior to each disaster's occurrence. In many of these cases and others like them, national and international governments and non-governmental organizations (NGOs) alike expended tremendous human and financial resources in an effort to promote public disaster preparedness. Clearly, most of these efforts fell far short of their expected outcomes.

Coupled with the recognition that natural and technological disasters are increasing both in number and severity, such poor community-wide public disaster preparedness figures obligate the adaptation of a more effective practice. Individual and family preparedness are vital to increasing overall community resilience, especially in light of the limitations typically experienced by the emergency services in the outset of large-scale events. Even in countries whose governments boast the most highly advanced emergency management capacities, leaders have found it necessary to warn average citizens that a minimum of 48 to 72 hours of self-reliance in the aftermath of a major disaster should be anticipated.

Individual emergency preparedness is by no means a new concept; however, the recognition of its true lifesaving potential has elevated its prominence among professionals in the field of emergency management. In response to a recent Council for Excellence in Government study, which reported that "most Americans haven't taken steps to prepare for a natural disaster, terrorist attack, or other emergency" (*USA Today*, 12/18/2006), former Department of Homeland Security Secretary Michael Chertoff echoed an even wider societal recognition of the dire need for increased public disaster preparedness efforts in stating that, "Everybody should have [disaster preparedness] basics down. I think Katrina shook people up. A lot of messaging and a lot of education, particularly at the local level, is the key" (*Government Executive*, 12/20/2006). Furthermore, the University of Colorado Natural Hazards Center, a leader in the advancement of emergency management throughout the world, stated in its January 2007 *Natural Hazards Observer* that: (1) there is a positive correlation between public awareness and positive disaster outcomes; (2) opportunities exist to better educate the public, coordinate messages, and initiate social change; (3) recent studies and surveys all indicate that there is an immediate need for better public education before disaster; and most importantly (4) there exists no comprehensive review of practices and resources and identification of components that make up an effective disaster public education program (*Natural Hazards Observer*, January 2007).

Perhaps most significantly, a June 2007 report released by the Emergency Preparedness Institute states, "The current approach to encouraging preparedness is ineffective, and a new method of communicating the importance of developing business and personal preparedness plans is needed." While other industries, most notably the public health sector, have enjoyed great success in shaping public attitudes and actions about their risk reduction behavior, the emergency management sector has thus far been largely unsuccessful in its endeavors. Despite the high cost and high profile of the penultimate preparedness effort, the Department of Homeland Security Ready.gov preparedness campaign has failed to make use of strongly supported public education methodologies that would have most certainly improved outcomes (*Washington Post*, 8/10/2006). Clearly, the most formidable obstacle to those preaching disaster preparedness is an industry-wide lack of knowledge about how people learn new behaviors, what influences them to act upon this knowledge, and the best way to create messages catering to those individual factors.

All communities are vulnerable to the effects of natural, technological, and intentional hazards. Every day, in every community, these hazard risks result in emergency events of varying size and intensity. Occasionally, they are of such great magnitude that they result in a major disaster. To minimize the consequences posed by known and unknown hazards, or to limit their likelihood of occurrence, communities perform mitigation and preparedness actions and activities. Individual members of the public, together representing the largest and most important community stakeholder, may be equipped with the skills and knowledge to further reduce their own, their family's, and their community's vulnerability if given the right kind of training using appropriate communication channels. This public, once prepared, becomes an integral part of the community's emergency management capacity. Properly trained individuals not only influence their own and their family's disaster risk, they also use the skills they have learned to rescue their neighbors, relieve shelter staff, retrofit homes for earthquakes, and to take countless more actions to extend the reach of their local emergency services.

As is true with the emergency manager and first responders in a community, members of the general public need information and training if they are to know what is best to do before, during, and after emergencies occur. The information provided must reflect their true risk and must be tailored to their needs, preferences, and abilities; transmitted in a way they can receive and understand; and tested for effectiveness. Any education provided will be received in conjunction or in competition with a wide range of other sources and messages relating to hazards, each considered "risk communication" regardless of its influence. In addition, while some of this coincident information will be accurate, effective, and useful, much of it is misleading, inaccurate, and ultimately harmful. Individuals are left to their own devices to cull through the daily onslaught of information received for that which will help them and their families prepare.

Creating risk messages and conveying them to the public require a dedication of time, planning, and learning. Unfortunately, informing the public about disaster preparedness is not as easy as simply telling them what they should do. The practice of disaster preparedness public education, which includes public awareness, education, and outreach, is an involved one relying on many years of practice and many different disciplines (including psychology, sociology, graphic design, marketing, communication, emergency management, and many others). Risk communication efforts are ongoing, long-term in nature, and must adhere to strategic plans to be effective. They should be coordinated with other providers in the community, and are most successful when they involve partners drawn from throughout the community and even beyond its borders.

This resource has been developed to provide practitioners in the United States and throughout the world, at both the local and national levels, with the background and the tools they need to plan, design, and carry out their public disaster preparedness efforts. The book is intended as an academic resource as well as a practical how-to guide.

ABOUT THE AUTHORS

Damon P. Coppola is the author of several leading emergency management academic and professional texts, including *Introduction to Emergency Management, Introduction to Homeland Security,* and *Introduction to International Emergency Management.* Mr. Coppola is also the co-author of several FEMA Emergency Management Institute publications, including *Hazards Risk Management, Emergency Management Case Studies, Comparative Emergency Management,* and *NIMS and Incident Management Systems.* As an independent consultant in the emergency management sector, Mr. Coppola has provided planning and technical assistance to emergency management organizations at the local, state, national, and international levels, and in both the nonprofit and private sectors. Mr. Coppola's clients have included FEMA, The World Bank, Save the Children, The Humane Society, ACORN, The Annie E. Casey Foundation, The Corporation for National and Community Service, Marriott International, and more. Mr. Coppola received his masters in Engineering Management (MEM) in crisis, disaster, and risk management from the George Washington University.

Erin K. Maloney holds a masters in communication from the University of Connecticut at Storrs, and is working toward her doctorate of philosophy in communication with a specialization in environmental science and policy at Michigan State University. Ms. Maloney's practical experience and research focus primarily on health communication, media, and research methods. Her research generally has implications for communication campaign message design and public processing of messages. Ms. Maloney consults on grassroots communication projects, most recently on a campaign that succeeded in passing a millage to preserve farmland and open space in the State of Michigan. Ms. Maloney is also an instructor and a research assistant in the department of communication at Michigan State University.

1

Public Disaster Preparedness

In Theory and in Practice

INTRODUCTION

Risk-related public education permeates all facets of modern life. Automobiles chime to remind us we must "Click It or Ticket;" cigarette packaging warns in no uncertain terms of the cancer risk contained within; Styrofoam coffee cup lids proclaim the obvious heat contained within; and medicine bottles are accompanied by pages of warnings and dangers to match a sentence or two of intended benefits. The flood of risk information we receive can be so great, in fact, that we simply stop paying attention to most of it (Figure 1.1).

Every moment of our lives entails risk. Moreover, for every hazard that threatens us there exist actions we may take by which our risk is increased or decreased. For many hazards, common sense would seem to dictate what constituted the wisest risk reduction measures: holding a handrail while descending a staircase; wearing a seatbelt while driving or riding in an automobile; avoiding cigarette smoke or quitting smoking. Unfortunately, such sensible advice is not always followed. Each year, in the United States alone, more than 1,600 people die by falling down the stairs; more than 23,000 perish in motor vehicle accidents while neglecting to wear a seatbelt (approximately 55% of all daytime and 66% of nighttime accident deaths); and approximately 440,000 people succumb to

FIGURE 1.1 Click It or Ticket Campaign — This image was developed by the National Highway Traffic Safety Administration for use by the media. (*Source*: http://www.nhtsa.gov/buckleup/ciot-planner/planner07/index.htm)

smoking-related illnesses (National Safety Council [NSC], 2007; National Highway Traffic Safety Administration [NHTSA], 2007; Centers for Disease Control and Prevention [CDC], 2003). Experience has demonstrated that, as a species, humans are not risk-averse and they often do not prepare for or even fear the right things. However, practice has also shown that risk-related misperceptions, miscalculations, and misguided behaviors can and have been corrected through the application of effective risk communication.

The majority of the risk-related public education we receive is generated by the public health sector. In fact, the most common avoidable or reducible risks we face as individuals fall under the domain of public health. For decades, public health professionals have studied the most common causes of death, discovered appropriate methods for reducing them, and developed effective messages and communication strategies to educate the public with this knowledge. These practitioners have steadily improved upon their public education methodologies, and their success rates in reducing population-wide risk have risen. Through public health education, people are living healthier, more productive, and longer lives.

Public health risks, however, form only one of many risk types we face. In fact, there are a great many larger-scale hazard risks that fall within the realm of an entirely different group of professionals: the emergency management community. Emergency managers and the various emergency services have been tasked with the heavy burden of preparing for, mitigating, responding to, and recovering from a full and growing list of natural, technological, and intentional hazards that each year affect millions of people worldwide and destroy billions of dollars in property, infrastructure, and personal and national wealth.

Emergency managers, like their counterparts in the public health sector, are acutely concerned with public risk. However, rather than addressing commonly occurring hazards that affect individual citizens on a more personal level — like heart disease and HIV infection — their foci have been those disaster-causing hazards that strike entire communities, cities, states, and even whole nations. In addition, unlike the public health sector, emergency management as a recognized profession is relatively new. Few emergency managers have the communication training of public health professionals. Fewer still have the practical experience required to develop and run a public education campaign. Moreover, almost no offices of emergency management have enjoyed means with which to adequately fund or even gauge the effectiveness of the campaigns they run.

The emergency management community is not entirely new to the public education arena. Its practitioners have long recognized the benefits of public education in terms of its ability to reduce population-wide risk from major hazards. Almost every American over the age of 40, for instance, possesses an instinctive understanding of the phrase "Duck and cover!" An even greater range of people understands what is meant by the command "Stop, drop, and roll." These phrases, developed to address the risk of an air raid in the first case and one's clothes catching fire in the second, are the products of two very widespread and successful emergency management public education campaigns institutionalized into the American school system (Figure 1.2).

Unfortunately, most emergency management public education efforts do not enjoy the widespread success that these two examples have. In fact, studies have shown that the vast majority of people do little or nothing to prepare for disasters and hazards, despite an increasing onslaught of information from local, state, and federal government agencies, the nonprofit sector, and elsewhere. The 21st century has thus far proven to be one marked by frequent and catastrophic hazards in the United States, including terrorist attacks, hurricanes, floods, and tornadoes. Despite extensive media coverage of these events, and what may be one of the

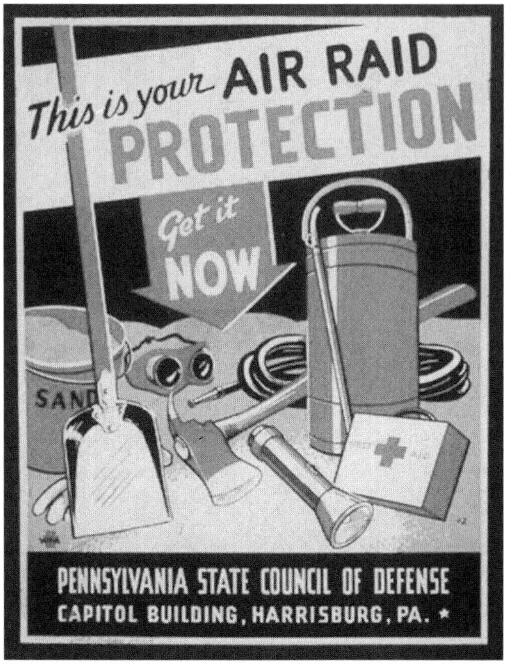

FIGURE 1.2 Civil defense era poster, Pennsylvania. (*Source:* Library of Congress, 2000.)

most widely touted emergency management public education efforts in decades (the Department of Homeland Security's Ready.gov Website and related Disaster Preparedness Month), recent research indicates that most individuals and families are still woefully unprepared for the risks they know to be affecting them (Council for Excellence in Government, 2007).

The poor success rates of the wider emergency management community are frustrating, but they in no way suggest that the goal of a "culture of disaster preparedness" is unattainable. Organizations like the American Red Cross, in fact, have illustrated through their CPR and first aid training programs that ordinary citizens can and are willing to learn how to help themselves and others in emergencies. The knowledge and experience of this organization that are attributable to its success are not widely enjoyed in the greater emergency management community. The public education work of the Red Cross has bridged the gap between public health and emergency management, and their practitioners have successfully incorporated the communication sector's lessons-learned into their public disaster preparedness education efforts.

Risk communication in practice is difficult at best, requiring a detailed understanding of the population targeted, the methods (channels) most suitable for reaching them, and the types of messages most likely to be received and acted upon. There is no such thing as a one-size-fits-all risk communication message, and any attempt to do such is doomed to fail. By learning and applying the effective practices developed over many decades by the public health community and others, emergency managers could enjoy similar levels of success.

This chapter introduces public disaster preparedness education and the concepts that guide its successful practice. To begin, a short overview of the experience gleaned in the public health sector, where risk communication efforts have advanced most significantly, is provided. This will help to familiarize readers with the relevant terms and concepts used by risk communication professionals, and the principles and theories that drive public education and preparedness behavior. Finally, the foundational elements of public education — its goals, limitations, and requirements — are presented and explained.

COMMUNICATION SCIENCE: A PRIMER

Communication science is a field of practice and research that has great potential to advance the efforts of emergency preparedness practitioners. This field is predicated on the fact that the mechanism by which information is conveyed to an individual or group plays an imminent role in the impact the message has on the intended message recipient. Many decades of research by this sector's experts have led to the discovery of six stages through which individuals process information, namely (McGuire, 1968):

1. Exposure to the message
2. Attention to the message
3. Comprehension of the arguments and conclusions presented in the message
4. Yielding to the message
5. Accepting the message
6. Information integration (which allows for message retention)

Thus, once individuals pay attention to and understand a message to which they have been exposed, they will use past experience with the issue to evaluate the new information. If after being compared to the old information the new information is accepted, it is integrated into one's knowledge structure. This integration is said to produce a change in one's belief system, leading individuals to change their attitude toward the topic. Attitudes are composed of three components: (1) cognitive (one's opinions or beliefs about the issue), (2) affective (one's feelings about the issue), and (3) behavioral (one's behaviors related to the issue). Because these components are so closely intertwined, influencing the cognitive and affective elements of an attitude may bring about behavior change.

The act of changing a person's attitude on an issue for the purpose of creating, reinforcing, or changing responses is called *persuasion*. Much of communication science and research seeks to uncover different techniques that can be used while delivering a message in order to make it more persuasive.

One of the most important factors practitioners should consider when attempting to construct messages to persuade an audience about an issue is the degree to which the members of the audience are involved with the topic being discussed. Involvement in this realm is defined as how much people feel that decisions about the issue will have direct implications for their lives and the things that are important to them; it is how much they care about decisions surrounding the issue. For example, some areas of the medical sector that may be considered high involvement issues are the legality of abortion and whether health insurance will cover the cost of prescribed medications. Some issues that might be considered low involvement issues are the importance of washing one's hands properly in order to prevent the spread of disease or whether health insurance covers the cost of a medication that you do not use.

Those who are highly involved with the topic are thought to scrutinize the message itself very carefully, so the degree to which they are persuaded by the message is based mostly on the strength of the argument presented. Therefore, for example, an individual with a great interest in emergency preparedness who is seeking information specifically on that topic will be likely to pay close attention to information presented in preparedness promotions. He or she will be likely to grasp upon inconsistencies and will question any counter-intuitive facts that are presented. If campaign planners are targeting an audience of highly involved people, it will be crucial for them to present logical, sound arguments. Also, if the communicator knows beforehand that the highly involved audience already holds an unfavorable attitude toward the behavior being presented, then he or she should be especially careful not to present any extreme viewpoints that counter the beliefs of the audience. For example, if a school board were already against the implementation of an emergency preparedness course into their students' regular curriculum because they think it would take away from other studies, it is better to acknowledge their viewpoint and work to find a compromise than to insist that there is no other option than the one being presented. Taking an extreme viewpoint that is counter-attitudinal to the audience has been shown to backfire in that the audience ends up spending the entire time counter-arguing (i.e., criticizing every point you make, finding something wrong with each of your points in their own minds, and ending up with a more extreme viewpoint against your message).

People who are uninvolved with the topic being promoted have been shown to pay less attention to the message itself. They will not likely be able to pick out inconsistencies in the message or counter-intuitive claims and instead will be influenced by features of the message (such as perceptions of source credibility, message length, and the sheer number of arguments in favor of the issues presented within the message). Therefore, a person who is not highly involved with the issue of emergency preparedness and is not seeking information about the topic is likely to be more influenced by a promotion that contains a list of semi-compelling reasons to engage in the behavior being promoted than a promotion that offers a single highly logical and rational reason.

Campaign designers seeking to promote emergency preparedness initiatives are very likely to find themselves dealing with audiences who are uninvolved with the issue. This can be positive in that they will not have strong attitudes that run counter to the promoted behavior, because they do not care as much about the issue. This can also be a drawback, however, in that no matter how logical and sensible it may be for people to engage in the behavior the communicator is promoting, it will likely take more than just a rational argument to get them to assume the inconvenience of a new behavior. Chapter 4 of this book discusses a number of different appeals and communication theories that have been successful in the past in persuading audiences and motivating behavior change. These appeals will be especially important for target audiences that are not involved with the topic of emergency preparedness. See Sidebar 1.1 for more information about the potential and limitations of communication campaigns.

Sidebar 1.1: The Potential and Limitations of Communication Campaigns

Communication campaigns are organized efforts that use communication to bring about a specified goal of informing, reinforcing, or persuading a defined group of people. While communication campaigns can be a very powerful tool of social influence, it is important to distinguish among the things that communication can and cannot do. In order to clarify these capabilities, the National Cancer Institute has offered the following (http://www.cancer.gov/pinkbook/page3):

Communication alone can:

- Increase the intended audience's knowledge and awareness of an issue, problem, or solution
- Influence perceptions, beliefs, and attitudes that may change social norms
- Prompt action
- Demonstrate or illustrate skills
- Reinforce knowledge, attitudes, or behavior
- Show the benefit of behavior change
- Advocate a position on an issue or policy
- Increase demand or support for social services
- Refute myths and misconceptions
- Strengthen organizational relationships

Communication combined with other strategies can:

- Cause sustained change in which an individual adopts and maintains a new behavior or an organization adopts and maintains a new policy direction
- Overcome barriers/systemic problems, such as insufficient access to healthcare

Communication cannot:

- Compensate for inadequate opportunities or services
- Produce sustained change in complex behaviors without the support of a larger program for change, including components addressing services, technology, and changes in regulations and policy
- Be equally effective in addressing all issues or relaying all messages because the topic or suggested behavior change

may be complex, because the intended audience may have preconceptions about the topic or message sender, or because the topic may be controversial

In addition to these limitations, it is crucial that practitioners recognize that successful communication campaigns are highly audience centered, emphasizing costs and benefits that are likely to produce an effect. Because of this, they generally focus on highly specified target audiences. In addition to narrowing the intended target audience, it is also important to remember that behavior change is quite an ambitious goal. It is not realistic to expect that people will change lifelong habits drastically as the result of even the most persuasive campaign. In other words, communication campaigns are effective when they attempt to persuade a specific group of people to change a very specific attitude or behavior; a single campaign cannot target "everyone" to do "everything" that needs to be done to solve a problem.

SOCIAL MARKETING

While theory is an essential component in developing messages that are most likely to inspire changes in behavior, the content of the message is only part of the persuasion process. Obviously, even perfectly constructed messages will have no impact if the intended audience is not exposed to them. As this text will demonstrate, message designers must have significant insight into the beliefs, attitudes, values, and current behaviors of their intended audience. Before messages are designed, practitioners must spend some time researching and gaining in-depth knowledge about the audience whose behavior they are trying to change.

Perhaps the most easily recognizable demonstration of the power of persuasion is the marketing of commercial products. Private companies spend millions of dollars on marketing each year and are generally confident that the capital dedicated to marketing will elicit a many-fold return in profit. In the past three decades, a number of nonprofit and pro-social sectors have begun to use marketing tools and techniques to circulate social change strategies more effectively. These strategies have guided the design and implementation of campaigns and interventions of successful governmental agencies and organizations such as the Centers for Disease Control (CDC), the U.S. Environmental Protection Agency (EPA), the National Institutes of Health (NIH), the U.S. Coast Guard, and the

National Traffic Safety Administration (NTSA), as well as nonprofit organizations such as the American Cancer Society, the American Red Cross, the American Heart Association, the American Lung Association, and the American Diabetes Association.

Social marketing campaigns employ the elements of the "marketing mix" used by traditional marketing firms. The traditional "four p's of marketing" include:

1. **Product**: In social marketing, the "product" is directly related to the end goal of the marketing campaign. The product may be a tangible good (e.g., hurricane straps), a behavior (e.g., clearing wildfire fuel, stockpiling food and water), a service (e.g., a home safety inspection), or an idea (e.g., family preparedness).
2. **Price**: The price refers to any cost (money, time, energy, embarrassment, etc.) associated with the product being promoted. It is essential to a marketing campaign to establish the perception among members of the target audience that the benefits of the product outweigh the costs associated with it.
3. **Place**: There are a number of different "places" that need to be considered in the design of a social marketing campaign. First, it must be determined where people may go to consume the product being promoted (e.g., where smoke detectors will be distributed, where response training may be offered, etc.). Practitioners also must decide upon placement of promotions in order to maximize exposure among the target population. This requires that promoters become familiar with the demographic being targeted, common media used among the group, and the ways in which members of this population obtain trusted information.
4. **Promotion**: Perhaps even more dependent upon practitioners' understanding of the target audience's beliefs, attitudes, and behaviors is the promotion itself. The promotion stage is the effort to communicate the message to the target audience. This stage takes the three previous "p's" into account in order to reach the target audience effectively with a message that promotes a clear product, emphasizing the benefits over the barriers of adhering to suggestions made by the message source.

In her widely employed book *Hands-On Social Marketing: A Step-by-Step Guide*, Nedra Kline Weinreich notes four supplementary "p's" of social marketing to be considered in addition to those used in the traditional marketing sector:

5. **Publics**: Weinreich draws a distinction between "external publics," which often include primary audiences (those most in need of influence) and secondary audiences (those who can be persuaded to contribute to influencing the primary audience), policymakers, gatekeepers, and "internal publics," composed of those who are involved with program implementation. Both are crucial to successful campaigns.
5. **Partnerships**: Issues being promoted by social marketing are often highly complex. Successful campaigns often involve multiple organizations playing different roles in the multifaceted goals of the campaign effort.
7. **Policy**: Sometimes the most realistic method of bringing about sustained behavior change is to implement policy change. Social marketing campaigns can aid in generating the support needed for such changes.
8. **Purse Strings**: Social marketing programs are often funded by grants and donations from foundations or the government. Practitioners should consider all funding possibilities before planning their social marketing efforts in order to keep their program plans within a realistic budget.

In a review of social marketing in public health, Walsh, Rudd, Moeykens, and Moloney (1993) noted that while the social marketing process differs from place to place, virtually all share in common that they are:

1. Disciplined in setting objectives and using a variety of techniques to achieve them
2. Centered around a target audience
3. Continuously refined throughout the campaign to meet the needs and desires of the intended audience

As these commonalities suggest, the necessity of constant research and adjustment according to the information and feedback provided by the target audience does not allow for a single sequential formula that all campaigns may follow to conduct an effective campaign.

The U.S. Department of Health and Human Services (HHS), NIH, and the National Cancer Institute have developed and refined a social marketing strategy known as the Health Communication Program Cycle. This tried and tested process has come to be known as a standard in the field of risk communication because of its comprehensive approach, elements of which have been highly effective in promoting preventive and reactive public health behaviors. Many of the elements developed for this process, and many of the associated actions, are easily transferable to the field of emergency management. The Health Communication Program Cycle divides the public health campaign process into four distinct stages:

Stage 1 involves planning and strategy development. It is essential that practitioners "do their homework" during this stage in order to better understand the problem and the role that communication can play in moving toward a solution. As noted earlier, there are no one-size-fits-all communication campaigns that inspire everybody to do everything. Therefore, while conducting this initial research, communicators are best served by identifying segments of the population who are in greatest need of an intervention. Based on this research, the communication team must then set realistic and measurable objectives involving a specific target audience, which are to be assessed at a given date. The research team must set goals that can be measured objectively throughout and at the end of the campaign to track progress and assess the effectiveness of the campaign.

Stage 2 involves developing and pre-testing concepts, messages, and materials for the campaign. During this stage, practitioners apply the insight gained from their problem (e.g., hazard risk vulnerability) and audience research previously conducted in order to create initial campaign messages and materials. A number of campaign messages and materials may be developed and pre-tested with focus groups, interviews, or surveys to get feedback from members of the intended audience. Pre-testing is used to eliminate all options except those that are best received by the audience they are intended to influence.

Stage 3 is the actual implementation of the program. In addition to reaching out to the target audience, communicators will track exposure and reactions to the campaign. These organized evaluations help to ensure that all materials are being distributed properly and to highlight aspects of the campaign that may need to be adjusted on an ongoing basis.

Stage 4 involves assessing the effectiveness of the campaign upon its completion and making refinements for possible future use. During this stage, campaign success is assessed by measuring how close communicators came to reaching each targeted goal. Many social scientists argue that campaigns without assessments are not worth conducting, as there are many documented examples of campaigns that produced either no effect or unintended negative effects that made the problem even worse. Campaign evaluation is discussed in greater depth in Chapter 5.

THE PUBLIC DISASTER PREPAREDNESS PROCESS: A SYSTEMS APPROACH

The systems approach presented in this text is drawn from the successes of the public health sector, where communication efforts have been conducted, studied, and improved upon for decades. This approach was developed to help planners ensure that they consider every important factor, identify every likely obstacle, and capitalize on all available resources, stakeholders, and partners who might support the effort. These methods are scalable to accommodate the scope of any project size, and may be used in whole or in part as the needs of the project dictate.

Public disaster preparedness education campaigns are complex endeavors, regardless of their size or scope. While it may be possible to hastily design, produce, and release individual risk information products and messages, these poorly aimed "hip shots" rarely achieve any measurable change in public behavior in the absence of a greater strategy. If what communicators seek is to facilitate actual improvements in community or population disaster resilience, they can only do so by ensuring their efforts are not performed in an ad hoc, disorganized manner or that they are not hastily planned without regard for each of the critical issues involved. To do either would be at best a waste of time and money, and at worst detrimental to the reputation of the organization and the safety of the target population.

All forms of risk communication depend upon an intimate understanding of the problems being addressed, the individuals and groups being communicated with, and the methods, partners, and resources available. The single most effective way to ensure that these factors are not only fully understood, but also appropriately utilized, is to follow a step-by-step systems approach.

We separate public disaster preparedness education approach into three separate components: (1) early planning, (2) developing a campaign strategy, and (3) implementing and evaluating the campaign. The following (Figure 1.3) is a brief summary of our approach.

FIGURE 1.3 Emergency management public education process diagram.

Phase 1: Early Planning

All projects must be built upon a solid, stable foundation of comprehensive planning. In this initial stage, communicators explicitly define the problem they are hoping to address, and perform an assessment to ensure that their estimations and impressions are as accurate as possible. At the same time, they define and analyze the target population, learning as much they can about their risk perceptions, abilities, wants, cultures, and other characteristics.

Once the problem has been defined and a target population has been selected, it is possible to conduct a thorough assessment of population risk. Based upon the outcome of that assessment, planners can begin to establish the goals and objectives of their campaign. They can also begin to form the larger planning team and communication coalition that will carry out the planning, management, and operations of the full campaign and that can identify, contact, and begin securing partners and stakeholders who will make the campaign possible through their assistance and resources. Finally, they begin to identify the various obstacles they may encounter in the course of their work, and predetermine ways in which to overcome them should they arise.

Phase 2: Developing a Campaign Strategy

The second phase in the process entails the development of a campaign strategy that guides actual operations after kickoff takes place, and the development of the materials and methods that are used to communicate with the target population. Communicators perform market and data research to determine what information already exists that either helps or hinders efforts, and determines how messages can best be transmitted. Using these data together with the analysis of the target population and an assessment of available resources, the communicators select channels, settings, and methods to execute their campaign. Message content and materials are designed and developed, followed by pilot testing to ensure the desired level of efficacy is attainable. Finally, activities and events are planned and staff is trained, all in preparation for the campaign that is about to begin.

Phase 3: Implementing and Evaluating the Campaign

It is in the third and final phase when communicators begin carrying out the various operational elements of their campaign strategy with the target population. During this phase, communicators regularly measure their levels of exposure and assess the success of their efforts. Using their findings, they evaluate and adjust their methods and materials to keep improving their ability to influence individual behavior. Communicators also measure how close they came to meeting or exceeding their goals. Using this information, they will be able to evaluate what actions and methods they chose that worked, and what they did or encountered along the way that hindered their progress. Project outcome evaluation is important for many reasons, as is described in Chapter 5.

PURPOSE, GOALS, AND OBJECTIVES

All disaster preparedness public education efforts share a common purpose, namely to:

> reduce individual vulnerability to one or more identified hazard risks as much as possible among as many members of a defined target population as possible.

Vulnerability, as will be explained in detail in Chapter 2, is the propensity to incur harm — in this case from the negative consequences of a disaster. There are countless methods by which vulnerability may be reduced and, likewise, resilience bolstered. However, the specific actions by which this is actually achieved — and to what degree of success it is achieved — are highly dependent on the ability of the preparedness campaign planning team to correctly identify the problem (i.e., the reason that vulnerability is high), assess the targeted population, identify the most appropriate methods to address the problem, and select the best mechanisms to communicate.

There are three primary goals of all public disaster preparedness education campaigns. While the most comprehensive campaigns might actually manage to accomplish all three of these goals, the majority often address only the first two.

These overarching public disaster preparedness education goals include:

1. Raising public awareness of the hazard risk(s)
2. Guiding public behavior, including:
   - Pre-disaster risk reduction behavior
   - Pre-disaster preparedness behavior

- Post-disaster response behavior
- Post-disaster recovery behavior

3. Warning the public

Goal 1: Raising Public Awareness of the Hazard Risk

The first goal of any public disaster preparedness education campaign is to notify the public about their exposure to a hazard risk and to give them an accurate impression of how that risk affects them personally. Because most people already have a general awareness that a hazard risk exists, this goal is most typically a simple matter of correcting inaccuracies and feelings of apathy toward preparedness for the particular hazard or hazards of concern. Adjusting public sentiment of this kind, which is most often the product of misguided assumptions regarding their need or ability to affect their fate, is likewise accomplished by raising awareness about the particulars of the hazards and risks of concern. While the occurrence of an actual disaster is the most likely and effective means by which people become aware of a particular hazard risk, it is preferable for obvious reasons that the public be enlightened long before a disaster happens. Additionally, the mere experience of surviving a disaster has not been shown to increase future preparedness behavior by any significant degree if a public disaster preparedness education effort does not follow the event (Citizen Corps, 2006).

The task of raising awareness involves much more than simply telling citizens what causes a particular risk. They must also be informed of how the risk affects them as individuals, what they are doing that places them at risk, and where and when the hazard will likely strike. They must fully understand the risk as it applies to them personally and to the population as a whole in order to effectively absorb, process, and act upon all subsequent information they receive.

Raising public awareness is a difficult task because of the competition communicators face for the attention of their audience. Communicators must always be aware that members of their target population face numerous risks on an individual level — many on a daily basis — that take up much of their limited attention. People already have to worry about their financial problems, crime, illness, problems at work and school, and many other immediate issues, before they even think about what the communicators are about to tell them. Leading risk communications expert M. Granger Morgan stated in a 2002 report on a study of how people receive disaster preparedness information that "the time that most people can devote to rare or unusual risks is usually very limited" (Morgan, Fischhoff, Bastrom, and Atman). In order to overcome these formidable social and psychological obstacles, public education aimed at raising hazard risk awareness must be as accurate, trustworthy, and effective as possible.

Goal 2: Guiding Public Behavior

Once an audience is informed sufficiently and appropriately about a hazard, they are primed to receive and process information that will help them take appropriate action to reduce their vulnerability to one or more hazard risks. This information will guide them in taking one or more of the following categories of risk reduction action (each distinguished by when the action is taken and for what greater purpose):

1. Pre-disaster risk reduction behavior
2. Pre-disaster preparedness behavior
3. Post-disaster response behavior
4. Post-disaster recovery behavior

Public education measures that address *pre-disaster risk reduction behavior* seek to instruct a population that is already aware of the existence of a hazard risk about the range of available options that can help reduce their individual and collective vulnerabilities to that risk. For instance, people living in areas where earthquakes are a problem might be shown how to secure their furniture to walls or floors to avoid the injuries that result when such items are toppled. Once informed about how their actions can affect their risk levels, people are more likely to act in ways that improve their chances of avoiding disaster in the future.

Public education aimed at teaching *pre-disaster preparedness behavior* attempts to inform the public about the kinds of actions they can take before a disaster happens that, while not necessarily reducing the likelihood of an event occurring, allow them to better manage the consequences they face. Actions taught in this category could include the stockpiling of certain materials; the establishment of individual, family, and community action plans, and the designation of appropriate and safe post-disaster meeting places.

Education in *post-disaster response behavior* seeks to teach an informed public how to react in the midst of and the aftermath of a hazard event. For instance, individuals must be instructed in how to recognize disaster indicators and know what they should do in response to them, including the proper way to participate in an evacuation. This category of public

FIGURE 1.4 Phuket, Thailand. Tsunami evacuation sign installed in a coastal tourist community hard-hit by the 2004 tsunami.

education also includes measures that empower the public to provide first-response services to their families, friends, neighbors, and themselves, supplementing the community's overextended emergency management resources (Figure 1.4).

Finally, education focused on *post-disaster recovery behavior,* which tends to be provided only in the aftermath of a disaster, teaches the disaster-affected members of the public how to best rebuild their lives. This can include helping people locate government, nonprofit, or international resources dedicated to relief and recovery, and learn how to provide those services for themselves.

Goal 3: Warning the Public

The final goal of disaster management public education is warning. Warnings are issued to alert an audience about a change in risk concerning an increased or certain likelihood of occurrence, and to provide them with authoritative instruction on appropriate actions they may take in response. Warning messages differ from awareness messages in that they instruct recipients to take immediate action. Like all risk communication efforts, the messages and systems developed to transmit a warning must be designed to reach the full range of possible recipients within the communities, regardless of location or time. Employing multiple systems, in collaboration with a full range of public, private, and non-governmental partners, increases the likelihood of a system reaching its target audience. Examples of the various groups that must be considered in planning for hazard warnings include individuals:

- At home
- In school
- At work
- In public spaces
- In their cars
- Who are disabled
- Who speak different languages
- Who are uneducated or have little education
- Who are poor

Warnings must inform people of an impending hazard or disaster and must instruct them on what to do before, during, and after the hazard (see Sidebar 1.2). Warnings may include information on how citizens can get more information, such as a Website, radio or TV station, or phone number. Public warnings are more than just a message. Warnings are built upon complex systems designed for the specifics of each hazard, population, and environment. Comprehensive warning systems seek to do most or all of the following, in order:

1. *Detect the presence of a hazard.* This step involves collecting data from a number of possible preestablished sensing and detection systems, including weather sensors, water flow sensors, seismicity and ground deformation sensors, air and water monitoring devices, and satellites, for example.
2. *Assess the threat posed by that hazard.* All hazards include some variable component of risk likelihood, which changes through time as more information becomes available. The data collected from the sensing and detection systems allow disaster managers

to update their assessments of the hazard and then consider how the community or country would be affected.

3. *Determine the population facing risk from that hazard.* The most effective warnings are those that target populations according to their risk, thereby ensuring that those not at risk avoid taking unnecessary actions, which can get in the way of disaster managers. Targeted warnings also allow responders to focus their assistance on those people with the most pressing needs.
4. *Inform the population.* One of the most difficult decisions disaster managers make is whether to issue a warning. Many are afraid that the public will panic if they are told about a disaster or that they will accuse the disaster manager of "crying wolf" if the hazard does not materialize. However, researchers have found both these outcomes to be rare in actual practice. Moreover, if the disaster management agency has followed established guidelines on risk assessment, their decision on issuing a warning can only be regarded as responsible.
5. *Determine appropriate protective actions that may be taken.* Using their updated assessment of the situation, disaster managers must determine which protective actions the public should be instructed to take. For some hazards, such as chemical releases, the public may have been told about multiple, conflicting actions, such as both evacuation and sheltering in place (remaining at their indoor location while sealing off the outside environment as much as possible).
6. *Direct the public to take those actions.* Through previous education efforts, the public should already be aware of the hazard and knowledgeable about the types of actions that may be required during a warning. Disaster managers must decide on the best course of action and relay that information to the public through previously established mechanisms. A warned public will seek information on what to do next, and it is important that a clear message be given to guide them.
7. *Support the actions being taken by the public.* Actual response assets (such as police and fire officials, emergency management officials, volunteers, and other established responders) should assist the public in following any broadcasted instructions; for instance, facilitating evacuation efforts.

Warning systems are much more than the application of technology and last-minute decisions. An effective warning system involves three distinct processes that are crucial so the public will actually take appropriate action. The three processes are:

1. *Planning.* During this first phase, disaster managers must consider what hazards allow for warnings, how and when the public will be warned, what the public can do in response to those warnings, what terminology will be used, and what authority and equipment are needed to issue the warnings.
2. *Public education.* The public will not automatically respond to a siren, announcement, or other form of warning just because the warning is given. Studies have shown that even with education about warnings, as few as 40% of recipients will take appropriate action. Without previous instruction on what to do, it can be assumed that even fewer would respond. A full explanation of warnings has to be incorporated into regular public disaster education campaigns, including what they will sound like, what they mean, where more information can be obtained, and the possible actions that will be taken in response to them.
3. *Testing and evaluation.* Testing and evaluation are necessary to ensure that recipients are not exposed to the warning process for the first time during a disaster. Testing allows citizens to experience the warning in a low-stress environment and to hear the actual sound or wording of a warning when they are neither anxious nor scared. Testing also allows disaster managers to ensure that their assumptions about the system and its processes reflect what will actually take place during a real warning event. Evaluation of the warning system helps to ensure in advance that the system is as effective as it can be.

Sidebar 1.2: The Words of Warning

Many different words are used to describe the severity of a hazard warning. Confusion about these words may cause recipients to respond with either too much or too little action. Therefore, clarity and consistency are important. In general, the terminology used to describe warnings includes:

- *Warning.* The hazardous event is under way or is highly likely to occur soon. Generally, an immediate threat to life and property exists. The public should take immediate protective action.
- *Advisory.* Advisories, like warnings, are given for events that are currently occurring or are about to occur. Advisories apply to events that are less severe than warnings in terms of the expected consequences to life and property. However, action to protect life and property are strongly recommended.
- *Watch.* Watches are issued when the likelihood of a hazardous event has increased significantly, but where and when the event will occur is uncertain. Watches are issued so that recipients may begin taking precautionary measures as far in advance as possible, even though there is a significant chance that the event may not materialize.
- *Outlook.* An outlook is a prediction of a hazardous event in the near future, based upon conditions that are beginning to look favorable for the onset of that event. Outlooks do not usually include action information or recommendations to prepare for the possible event.
- *Statement.* Statements are not warnings themselves, but are used to provide detailed follow-up information to warnings, advisories, or watches.

Other Risk Communication Goals

Like all other forms of public education, public disaster preparedness education seeks to accomplish several goals simultaneously, often in addition to the three primary goals listed previously. Goals serve to help planners focus their efforts and ultimately determine if their campaign was successful in achieving what it set out to do. Goals should be basic, should be attainable, and should complement one another. They should also suit the communication theory and practice employed in the campaign. Through the actions that are taken to meet these goals, public knowledge and skill are increased and resilience results. Additional goals that address specific needs of an identified and assessed problem, a profiled population, and a developed message (as identified by the CDC [1995] and other sources [Baker, 1990]) are as follows:

1. Increase or enhance knowledge.
2. Refute myths/misconceptions.
3. Influence attitudes and social norms.
4. Develop skills.
5. Reinforce knowledge, attitudes, and behaviors.
6. Suggest/enable action.
7. Show the benefits of a behavior.
8. Increase support or demand for services.
9. Coalesce organizational relationships.

Each of these goals is relevant to the campaign, and therefore important, only if it contributes directly to what the campaign is ultimately seeking to achieve, namely:

- Hazard awareness and accuracy of information
- Behavior change
- Individual and population-wide resilience (i.e., vulnerability reduction)
- Sustainability

Priorities and Goals of Risk Communication Recipients

Risk exists in many forms, as experienced and perceived by individuals. In a study performed by risk communication expert M. Granger Morgan and his colleagues, citizens were asked to make lists of risks that concern them the most. The responses ranged from threats that would result in injury or death, such as accidents, disease, and crime; to economic risks that would result in a financial loss for the individual; to personal concern-related risks, such as love-life problems or problems in school or at work, among others. Only 10% of the risks that were cited related to natural or technological hazards. Morgan and his team found that "whereas professional risk experts devote many hours to considering rare and unusual hazards, most people do not share this preoccupation. With jobs, family, friends, and the other demands of daily living, their lives are filled with more immediate concerns." The authors summarize this statement in saying, "The time that most people can devote to rare or unusual risks is usually very limited" (Morgan et al., 2002 , p. 2).

For some of the risks people must face, such as those associated with health (e.g., smoking) or safety (e.g., driving), there is a certain amount of control that individuals possess to directly minimize their risk. For other risks, people can only indirectly minimize their risks, through social processes (i.e., voting for certain risk controls or mitigation measures). However, the authors write that

> In all cases, [people] need a diverse set of cognitive, social, and emotional skills in order to understand the information that they receive, interpret its relevance for their lives and communities, and articulate their views to others. They can acquire those skills through formal education, self-study, and personal experience. However, as diligent as they might be, individuals are helpless without trustworthy, comprehensible information about specific risks. (p. 2)

Target populations, like risk communicators, have their own goals that dictate their participation in a public education effort. While there will always be a certain percentage of the target population that pays no attention at all to communicators or their messages, most will recognize the pertinence of the messages to their own life and seek more information upon which they will determine future action. How they go about this and what kind of information they seek will not be uniform across all issues they face. Once they elect to pay attention to the communicator's message, there are generally three areas that describe their intentions for doing so.

1. **Advice and Answers** — This group consists of people who want only to be told exactly what to do. They need step-by-step instruction on how to address the problem of which they have just been informed. This group of people would rather be given the product of an analysis process than be given all of the necessary information to draw conclusions themselves. This can be compared to the faith that is held in the advice given by a trusted doctor, lawyer, or other consultant.
2. **Numbers** — There are certain people who will not take advice at face value. Rather, this group would prefer the option to draw their own conclusions about what they need to do from a body of statistical evidence that is provided to them by communicators. This group would rather be faced with quantitative summaries of expert knowledge than to be fed step-by-step instruction offering little room for individual analysis and perspective.
3. **Process and Framing** — This group is composed of those individuals who prefer to fully analyze as many factors relevant to the problem as possible, assuming full ownership of their actions and any

likely outcomes that result. Members of this group generally require three primary data points with which to shape future action:

i. The costs associated with inaction versus those associated with the actions prescribed by communicators
ii. The statistical likelihood of falling victim to a disaster and the probable consequences should victimization occur
iii. The actual or expected reduction in risk that would be achieved through each of the actions offered

This group is the most motivated and capable of the three, and is interested in learning not only how vulnerabilities are created through their own or their community's actions, but also how they or their community can control that risk. Using that information, they understand that they will be able to assess their personal circumstances, make individual identifications and assessments of risk, and devise self-tailored solutions. (See Figure 1.5.)

Communicators must also be aware that the priorities and goals of their target audience may not closely match their own. Many individuals' goals may be directly opposed to the communicators', or even opposed to risk reduction in general. For instance, individuals who seek a waterfront view for their home may elect to place themselves in a location that also provides the most direct exposure to the forces of hurricanes, storm surges, and erosion. While the ultimate reduction in risk would be to remove such an individual from the source of risk (the oceanfront property), such an action would probably be in too great opposition to their wants and needs to bring about behavioral change. By identifying and accommodating the goals of recipients, and developing and communicating strategies that offer multiple solutions, communicators will achieve much greater success overall (even if the prescribed solutions would not have achieved as significant risk reduction under more perfect circumstances).

Sidebar 1.3: The Various Publics

Public disaster preparedness education can be a powerful tool in helping not only individuals to change, but also groups, communities, and even whole societies. The following list, compiled by the NIH, illustrates the different ways that risk communication can affect different segments of society in an effort to bring about change:

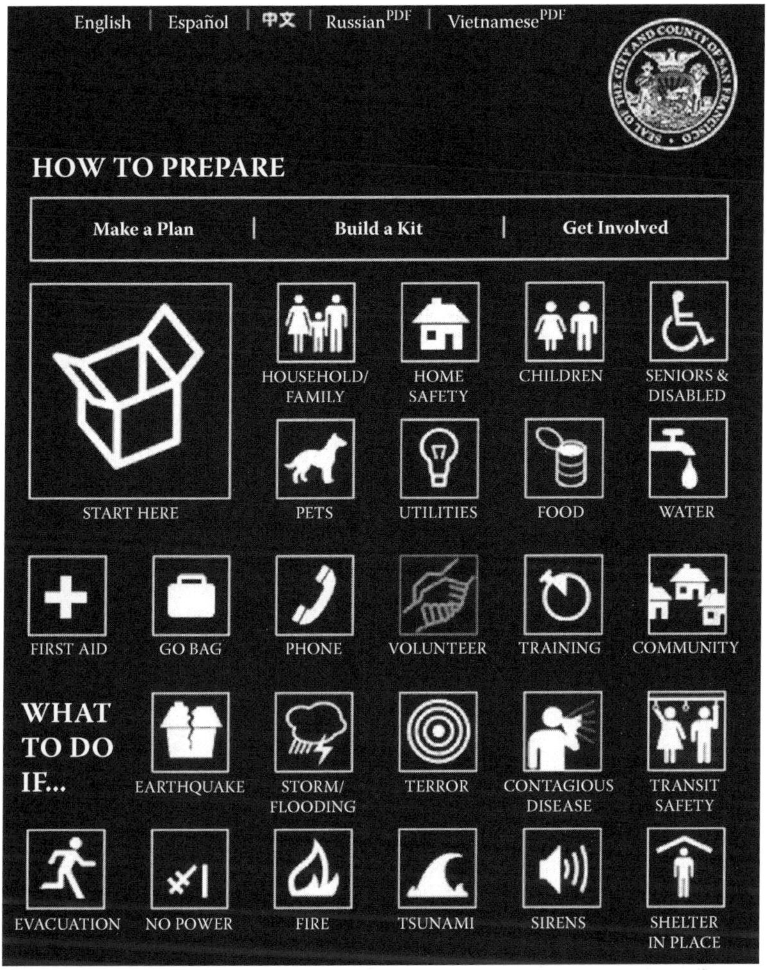

FIGURE 1.5 Screenshot from www.72hours.org, a public emergency preparedness education Website developed and maintained by the city and county of San Francisco, California.

- **Individuals** — The interpersonal level is the most fundamental level of risk-related communication because individual behavior most profoundly affects individual vulnerability. Communication can affect individuals' awareness, knowledge, attitudes, self-efficacy, skills, and commitment to behavior change. Activities directed at other intended audiences for change may also affect individual change, such as training employees in how to apply at-work preparedness measures to their home life.

- **Groups** — The informal groups to which people belong and the community settings they frequent can have a significant impact on their personal vulnerability and propensity to participate in preparedness behavior. The groups include relationships between customers and employees at a salon or restaurant, members of the same gym or social club, students and parents in a school setting, or employees at the same work site. Risk communication is easily passed from a direct recipient to others within his or her informal social networks. In fact, activities that are aimed at informal settings often enjoy much greater success than those relying solely upon direct formal communicator-to-recipient patterns.
- **Organizations** — Organizations are groups with defined structures, such as associations, clubs, or civic groups. This category also includes businesses and government agencies. Organizations can carry preparedness messages to their constituents, provide support for preparedness communication programs, and make policy changes that encourage individual change.
- **Communities** — Community leaders and policymakers can be effective allies in influencing change in policies, products, and services that can hinder or support people's actions. By influencing communities, disaster preparedness communication efforts can promote increased awareness of existing hazards, changes in attitudes and beliefs, and group or institutional support for desirable behaviors. In addition, communication can advocate policy or structural changes in the community (e.g., floodplain management and mitigation grant funding) that encourage resilient behavior.
- **Society** — Society as a whole influences individual behavior by affecting norms and values, attitudes and opinions, and laws and policies, and by creating physical, economic, cultural, and information environments. Public disaster preparedness education aimed at the societal level can change individual attitudes or behavior and thus change social norms. For instance, greater awareness of the terrorism hazard brought about by

the September 11th terrorist attacks resulted in public support for much greater funding for local first responders.

Source: United States Department of Health and Human Services (HHS). 2002. *Making Health Communications Programs Work.* Bethesda, MD: National Institutes of Health.

SEEING THE BIGGER PICTURE: COMMUNICATION AS ONE COMPONENT OF A LARGER SOLUTION

Public disaster preparedness education campaigns are conducted to address existing hazard vulnerabilities. Like all public education efforts, when properly conducted, they can be highly effective in limiting behavior that fosters vulnerability and promoting behavior that reduces it. Conducted in a vacuum, however, these efforts rarely succeed. This is primarily because vulnerabilities tend to have much deeper social roots that extend far beyond a simple lack of knowledge (Paton and Johnson, 2001).

To maximize the effectiveness of the public disaster preparedness education campaign, planners will need to view it as one attack in a battle that is being conducted on several fronts. In the planning process, as is detailed in Chapter 3 and Chapter 4, the roots of the vulnerability must be assessed. These social factors may be financial, cultural, religious, or of other origin. Likewise, as is true with all social problems (e.g., crime, health, unemployment, homelessness), communication programs should be integrated as much as possible into any larger, comprehensive efforts that are currently addressing or that plan to address the same problem from a similar or different perspective.

The various available components that are typically incorporated into a comprehensive public disaster preparedness campaign include:

- **Communication** — Communication, as subsequent chapters illustrate, can come in many forms and by many different methods. The target of these communication messages includes not only the vulnerable population, but also the general public, educators, the media, emergency services, non-governmental organizations (NGOs), policymakers, or anyone else who either needs to make a behavioral change or can facilitate one.

- **Facilitation** — Facilitation involves helping members of the target population take necessary action to reduce their vulnerability. Oftentimes facilitation is necessary because certain individuals are able to receive and understand the messages communicators are sending them, but are unable to act upon them without outside assistance. Facilitation can also be a way to prompt reluctant individuals who are otherwise able to take action to do so. Facilitation differs from communication in that the facilitators are actually performing or assisting others in performing the tasks required to change vulnerable behavior.
- **Funding and financial incentives** — While not all vulnerability reduction solutions offered by public disaster preparedness education campaigns have an accompanied cost for recipients, most do. By providing a grant or other financial incentive program (like tax breaks or insurance discounts) to the target population, it is often possible to prompt those for whom only money is a restrictive factor into action. In many cases, even for those individuals who can afford the risk reduction measures but choose not to out of ambivalence or some other self-limiting factor, the chance to gain something for free or at a significantly reduced cost is enough to initiate action.
- **Policy change** — Public policy, which includes laws, standards, regulations, or operating procedures, for example, has long been an effective way to create behavioral change. Policy action is best suited in several different types of situations, such as when social barriers prevent individuals who would otherwise prepare for a hazard from doing so (e.g., requiring the posting of evacuation signs in hotels), when a recognized safety standard is known but not yet required in private life or on private property (e.g., building codes, seat-belt laws), or when implementing such laws or requirements would bring greater benefit to the greater good of the public (e.g., obligatory first aid education in public school).
- **Technology** — Technology alone is rarely an answer in the emergency management world, but by harnessing technological advances and devices, emergency managers have been able to significantly increase their ability to reduce death and destruction among the affected population. This also applies to public disaster preparedness efforts, in that technology can help break down existing barriers and improve the effectiveness of other efforts.

For instance, various monitoring and detection systems have allowed more accurate mapping of hazard risk, which in turn has allowed people living or working within those risk zones to better understand their exposure. Technology research has also helped to develop cheaper and easier-to-use mitigation and preparedness devices that allow the public to reduce risk on a personal level. Widespread use of the National Oceanic and Atmospheric Administration (NOAA) weather radio is a perfect example of how technology has improved preparedness (see Sidebar 1.4).

Sidebar 1.4: NOAA Weather Radios

NOAA Weather Radio All Hazards (NWR) is a nationwide network of radio stations broadcasting continuous weather information directly from the nearest National Weather Service office. NWR broadcasts official Weather Service warnings, watches, forecasts, and other hazard information 24 hours a day, 7 days a week. Working with the Federal Communication Commission's (FCC) Emergency Alert System, NWR is an "All Hazards" radio network, making it the single source for comprehensive weather and emergency information. In conjunction with federal, state, and local emergency managers and other public officials, NWR also broadcasts warning and post-event information for all types of hazards, including natural (such as earthquakes or avalanches), environmental (such as chemical releases or oil spills), and public safety (such as AMBER alerts or 911 telephone outages).

Known as the "Voice of NOAA's National Weather Service," NWR is provided as a public service by NOAA, part of the Department of Commerce. NWR includes more than 985 transmitters, covering all 50 states, adjacent coastal waters, Puerto Rico, the U.S. Virgin Islands, and the U.S. Pacific Territories. NWR requires a special radio receiver or scanner capable of picking up the signal. Broadcasts are found on the VHF public service band at seven frequencies between 162.400 and 162.550 MHz.

NOAA broadcasts cannot be heard on a simple AM/FM radio receiver. However, many receiver options range from handheld portable units that pick up only NWR to desktop and console models that receive NWR in addition to other broadcasts. The National Weather Service does not manufacture, sell, or endorse any particular make or model of receiver. Receivers can be found at many retail outlets, including electronics, department, sporting goods, and boat and marine accessory stores and their catalogs. They can also be purchased via the Internet from online retailers or directly from manufacturers. Many municipal mitigation and preparedness programs have used grant funding to supply residents with receivers free of charge, in conjunction with training in how to operate them. The following are types of NWR receivers:

- **Standalone:** While these receivers might also come with AM/FM bands, their primary use is to receive NWR broadcasts. Buyers can choose between handheld and desktop models, depending on the portability desired. Prices range from around $20 to over $100.
- **Multi-band/function:** These receivers bundle a number of features, and NWR is just one of possibly many frequency bands included. This includes AM/FM radios, shortwave receivers, CB radios, VHF marine radios, scanners, and more.

The following are features typical of NWR receivers:

- **Tone alarm:** The National Weather Service will send a 1050-Hz tone alarm before most warning and many watch messages are broadcast. The tone will activate all the receivers that are equipped to receive it, even if the audio is turned off. This is especially useful for warnings that occur during the night when most people are asleep.
- **SAME technology:** SAME, or specific alert message encoding, allows users to specify the particular area for which they wish to receive alerts. Most warnings and watches broadcast over NWR are county based or independent city-based (parish based in Louisiana), although in a few areas of the country the alerts are issued for portions of counties. Since most NWR transmitters are broadcasting for a number of counties, SAME receivers will respond only to alerts issued for the area (or areas) selected. This minimizes the number of "false alarms" for events that might be a few counties away from where users live.
- **Selectable alerting of events:** While SAME allows users to specify a particular area of interest, some receivers allow users to turn off the alarm for certain events that might not be

important to them. For example, if you live in a coastal county, but not right at the beach, you might not care about coastal flood warnings.
- **Battery backup:** Since power outages often occur during storms, having a receiver with battery backup can be crucial.
- **External antenna jack:** While most receivers come with a whip antenna that can usually be extended out from the unit, depending on location users may need an external antenna to get good reception. Some receivers come with an external antenna jack that allows users to connect to a larger antenna.

Source: NOAA, 2008.

Unfortunately, a great many public disaster preparedness education campaigns are conducted wholly independent of any other supportive approach that would likely have increased the desired change significantly. Without such support, campaign planners set unrealistic expectations for themselves concerning that which they can ultimately accomplish. In most cases, the non-communication components of a preparedness campaign are already being conducted or are available to recipients; however, the communication campaign planners are simply not aware of them and are therefore unable to link their efforts. Thorough planning in the design phase can help to locate as many of these other options as possible, thereby incorporating them into the messages and actions that are developed and delivered.

Consider a campaign to increase the use of household smoke detectors to prevent injuries and deaths from fires (which cause over 4000 fatalities and tens of thousands of injuries each year in the United States alone). Through effective risk communication, residents are informed that they are at risk from house fires, told how that risk applies to them, and given appropriate information about the value the use of a smoke detector could provide in terms of giving them early warning about a fire in their house. Even if translated into all applicable languages and broadcast on all appropriate channels, using all necessary communication campaign components (as described in this text), this communication effort alone is likely to do little more than raise greater awareness of the hazard. Such a limited campaign is unlikely to increase the number of homes that use smoke detectors significantly above rates that were seen prior to the campaign. By understanding why people might not be using the devices, and what else can be done to increase their use through the application of a more comprehensive preparedness campaign, it is possible to achieve much greater results. For instance, it is often the case that there exist residents who, for one reason or another, lack the financial means to purchase the devices. Through a simple need-based grant program, communicators can provide additional information about how to acquire free smoke detectors, thereby eliminating the financial burden. Another obstacle might originate from a lack of residents' ability or time to install smoke detectors in their homes. By promoting or creating a home inspection and smoke detector installation program, especially one that carries no associated cost, another major roadblock will have been eliminated. By advocating or calling attention to local laws that mandate the use of smoke detectors in the home, it may be possible to reach another group — those who are reluctant to install smoke detectors for any number of reasons. Finally, by promoting the many technological advances that have been made in the smoke detector industry, including systems that contact the fire department directly when triggered and others that are able to alert people who are deaf or are sleeping, vulnerability is decreased even further. (See Figure 1.6.)

- To see an example of a program that offers smoke detectors free of charge, managed by the James City, Virginia Fire Department, visit: http://www.james-city.va.us/fire/smoke-detector.html
- To see an example of a program that offers funding for smoke detectors for hearing impaired citizens, managed by the Ohio State Government, visit: http://olrs.ohio.gov/asp/pub_SmokeDetectorHandbook.asp
- To see an example of legislative action taken to drive the use of smoke detectors, passed in Baltimore City, Maryland, visit: http://www.peoples-law.org/housing/ltenant/legal%20info/smoke%20detectors.htm

---

The key to ensuring a comprehensive campaign is not necessarily carrying out each of the components discussed previously. Rather, it is contingent upon the communication campaign planners to recognize these other non-communication needs, to identify existing or likely sources to accommodate them, and to incorporate them into the overall preparedness effort. This is described in detail in Chapter 4.

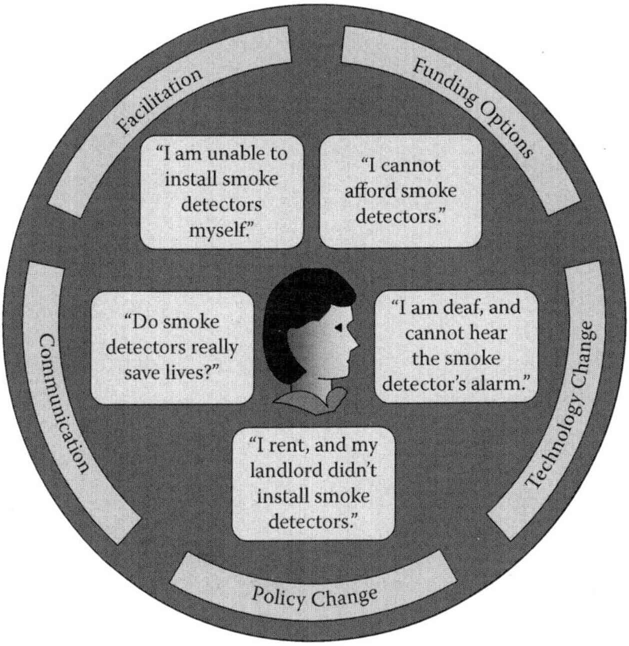

FIGURE 1.6 A comprehensive smoke detector safety campaign.

REQUIREMENTS OF A PUBLIC EDUCATION CAMPAIGN

Although there is no single recipe by which all public education campaigns are developed, there do exist essential ingredients without which success will range from difficult to nearly impossible. Perhaps the most obvious requirement is that of trust in the communicator. Recipients of risk information are unlikely to heed any instructions they hear or read if they cannot lend any credibility to the source of those instructions. However, trust only opens the door — a full range of components factor into how well received a message ultimately finds itself among the target population. Consider the characteristics of an effective public education message that are listed in the following, each of which has been identified and verified through years of research and practice in the public health communication field:

- **Trustworthy** — Public disaster preparedness education efforts ask people to alter their behavior. This is a very personal request, and one for which a great deal of trust is involved. If communicators cannot be believed, few will follow their instructions. One of the most effective means of establishing trust is to indicate a connection with an organization or institution that is highly regarded within the community or by members of the population, or by having popular community representatives (especially from within the target population) perceived as being involved in the communication effort.
- **Authoritative** — Individuals are free to choose to whom they listen. Even if trust has been established, people need to believe that a communicator is an authoritative source of information, or else they are likely to hold the views of another source (which may be opposing or conflicting) more favorably. Recipients of the message must have confidence that the communicator is qualified and sufficiently informed to tell them what to do.
- **Free of personal gain** — In a capitalist society, people receive a barrage of advertising and consumerist information on products and services. There is an inherent skepticism among the public that nearly everyone is out for personal gain. Communicators must be able to convince the public that their message is conveyed only to serve the public good, not to gain anything in return. This can be difficult to do, especially in instances where preparedness requires recipients to purchase something or otherwise spend their money.
- **Accurate** — Communicators must always assume that the recipient public is going to verify any information that is communicated to them prior to acting on that information. For this very reason, it is vital that any statistical or factual information be highly accurate and as current as possible. For specific statistical information, it is often wise to cite a credible reference to back up whatever claim is being made.
- **Consistent** — Public education campaigns are rarely just a one-time broadcast of a message. Rather, the campaign will seek to repeat the message to ensure that it is received over a wide enough segment of the target population and that it becomes memorable to those who hear it. If the message that is transmitted is not the same over time, the benefit of repetitiveness will be lost. This is especially true if a catch phrase is used (e.g., "Stop, drop,

and roll"). If the main point or argument of the message changes, it will seem less credible to the recipient audience. This same requirement extends to the use of multiple channels. For instance, a message that is transmitted over the radio should match messages in print, on the Internet, and elsewhere, or else recipients who make contact with multiple channels will likely become confused. Finally, if different speakers present the message, each must be consistent with the others, even if they are speaking to multiple distinct target populations.

- **Repetitive** — As the previous requirement suggests, recipients must face repetitive exposure to a message before they are able to learn from it. As is true with product marketing, the more exposure to a message that a person has, the more memorable that message becomes.
- **Easily understood** — Naturally, message recipients must be able to understand a message if they are to receive it. There are many components to this requirement. First, it must be in a language they speak fluently or well enough to understand correctly what is being said. Second, it must match their level of education if it is to be fully understood (which includes writing using more technical language for more educated populations as well). It should be relevant to situations and scenarios that are understandable to the audience as a gender, culture, age, geographic grouping, and other demographic measures.
- **Rational** — In addition to a message being understood, it must make sense in a rational and logical sense. Arguments have to be as simple as possible for the audience targeted, and defensible justifications must be presented to support the measures advocated. They should also contain concrete information on the addressed hazards, including mortality rates, affected area, hazard time frame, population affected, and others.
- **Accessible** — All people receive their information in a unique manner, based upon their preferences, access, social networks, and other factors. Communicators must ensure that, within this realm of information reception, the message is easily accessible to the audience being targeted. This may require multiple channels if the population is a diverse group or if their patterns of learning dictate such measures. More information on channels is provided in Chapter 4.

- **Solution-oriented** — To be truly effective, a communication message must not only inform members of the target population about their hazard risk, must also inform them of realistic and affordable solutions by which they may reduce their personal vulnerability. This was one of the greatest early criticisms of the Homeland Security Advisory System, which uses five color codes to indicate terrorism hazard risk. Devoid of any real behavioral modifications or solutions to individual vulnerability, the warnings did little to adjust preparedness behavior among the targeted general audience.

Sidebar 1.5: Characteristics of an Effective Campaign

Public disaster preparedness campaigns have specific requirements without which most efforts would fail. In addition to these requirements, there are certain characteristics that, while not a requirement, certainly do lend themselves to the success of the campaign. The University of Wisconsin (1995) developed the following list of several of these characteristics:

- Long-term rather than short-term.
- Built-in evaluation of effectiveness in the mid-term.
- Incorporate the educational infrastructure, from primary schools to university level.
- Use the mass media to reach the majority of the audience.
- Use folk or alternative media to reach other sections of the population.
- Aim at both the general public and the special publics (i.e., minorities in the population).
- Aim at official and quasi-official levels.
- At first, begin in an area with recurring disasters or hazardous conditions, where there is some incentive for the population to react favorably to the program.
- Later, establish a truly national program welded into the educational and communicational aspects of national development.
- The program should be national, regional, or local in character to increase the motivation of those taking part in the program, as well as their knowledge about the hazards.

- It should be an extension of the public information program.
- It should take advantage of normal conditions of life, normal modes of behavior, normal sources of information, and so forth in the preliminary stages.

THE DANGERS OF FAILED RISK COMMUNICATION

Individuals and organizations embarking upon a risk communication effort are assuming a great responsibility with many moral implications. They are asking their target audience to trust that the content of their messages will improve their safety and security. They are creating a working relationship in which the recipient becomes willing to dedicate limited time and resources to the advice of the communicator. For this reason, communicators must be certain that their messages and strategies are accurate and effective not only in a general sense, but in a specific sense as well, regarding the particulars of the target audience. Communicators must be sure that their target audience easily understands the messages and channels through which they communicate risks to the public. Such high standards dictate the need for testing prior to widespread application, and testing and adjustment during the course of the campaign, as described in Chapter 5. The risk of neglecting this requirement is too great, as failed communications can result in any or all the following:

- Waste of the target audience's time and resources
- Waste of resources dedicated to the risk communication effort
- Loss of empowerment for individuals seeking to manage their risk
- Resentment toward the communicators or authority figures
- Understanding the message, the problem, and the recommended solution
- Inadvertent controversy and conflict
- Creations of threats larger than those posed by the risks they intended to address
- Undermining of effective decision making
- Increased confusion and misunderstanding about hazard risk
- Increase vulnerabilities that it sought to decrease

Such failures are not always deliberate, as they can result from the failure of communicators to grasp the complexity of the message they are attempting to convey to a lay public.

CONCLUSION

Risk communicators embarking on a public disaster preparedness education campaign must understand that they are assuming a difficult task with significant consequences. However, thanks to the research and practice obtained by professionals in the emergency management, public health, and communication science fields, they need not go into these efforts blind. By following a systemic approach, like the one presented in this chapter and throughout this book, risk communicators can ensure that they are making every possible effort to understand the problem at hand, study their target population, design their messages correctly, select appropriate channels, test and pre-test their methods, and ultimately cause the change that brings about disaster resilience.

REFERENCES

Baker, Frank. 1990. Risk communication about environmental hazards. *Journal of Public Health Policy*. 11(3), 341–359. Autumn.

Centers for Disease Control and Prevention (CDC). 2003. Cigarette smoking-attributable morbidity — United States 2000. *MMWR Weekly*. 52(35), 842–844. September 5.

Centers for Disease Control and Prevention (CDC). 1995. *Guidelines for Health Education and Risk Reduction Activities*. Atlanta, GA: Author.

Citizen Corps. 2006. Patterns in current research and future research opportunities. *Citizen Preparedness Review*. Issue 3. Summer.

Council for Excellence in Government. 2007. America's Preparedness for Disaster or Emergency Improves; Nation's "RQ" Score up to 4.1. Press Release. October 1.

McGuire, W. J. 1968. Personality and attitude change: An information processing theory. In *Psychological Foundations of Attitudes*, A. G. Greenwald, T. C. Brock, and T. M. Ostrom, Eds., San Diego, CA: Academic Press, pp. 171–196.

Morgan, M. Granger, Baruch Fischhoff, Ann Bostrom, and Cynthia J. Atman. 2002. *Risk Communication: A Mental Models Approach*. Cambridge: Cambridge University Press.

National Highway Traffic Safety Administration (NHTSA). 2007. Motor Vehicle Traffic Crash Fatality Counts and Estimates of People Injured for 2006. http://www.nhtsa.dot.gov/portal/nhtsa_static_file_downloader.jsp?file=/staticfiles/DOT/NHTSA/NCSA/Content/PDF/810837.pdf

NOAA. 2008. NWR Receiver Consumer Information. NOAA Website. http://www.nws.noaa.gov/nwr/nwrrcvr.htm

Paton, Douglas, and David Johnson. 2001. Disasters and communities: Vulnerability, resilience, and preparedness. *Disaster Prevention and Management.* 10(4), 270–277.

United States Department of Health and Human Services (HHS). 2002. *Making Health Communications Programs Work.* Bethesda, MD: National Institutes of Health.

2

Managing Risk, Emergencies, and Disasters

INTRODUCTION

Disasters have maintained a constant presence throughout our historical record. In response, individuals and societies alike have struggled to reduce their exposure to the consequences of these disasters in order to maximize their likelihood of survival. They have taken countless steps to reduce the chances that disasters occur, developed measures to respond to the impact should they strike, and discovered ways to manage post-disaster relief and recovery needs of victims in the aftermath of these events. All of these efforts have sought a common goal: disaster management.

As civilizations evolved and governments assumed increasingly greater stewardship over the hazards that threatened their constituents, formal disaster management structures emerged. In contemporary society, almost every nation's government operates a unique office or agency dedicated to preventing hazards from becoming realized disasters, and managing their consequences when they do. Many have gone as far as implementing a statutory structure through which the management of hazards is standardized, regulated, and perpetually funded. Several nations' governments, including that of the United States, have even allowed for full decentralization of this emergency management capacity, thereby ensuring that each local government is able to tailor management needs to the specific hazards that exist at the community level.

Through science, technology, and innovation, nations have further contributed to an overall increase in human disaster resilience. The development of new and improved mitigation techniques has helped to control hazards by decreasing their chances of occurring or minimizing their consequences should they occur. Hazard detection and warning systems are annually more comprehensive in their ability to inform us of coming events hours and days earlier. These and other modern response tools, coupled with increasingly advanced and coordinated response systems, have helped to ensure that more and more people caught in the path of disaster survive.

Despite all that has been achieved, significant risk remains. Each year, millions of people fall victim to disasters, of which hundreds of thousands lose their lives. Earthquakes, floods, hurricanes, cyclones, tornadoes, landslides, terrorist attacks, and many other hazards that plague us will continue to cause financial and physical ruin due to our propensity to live where they strike — along coastlines, over faults, below volcanoes, and more. Moreover, with climate change altering the global hazard playing field, we may even see a reversal in our positive trend toward hazard control (see Figure 2.1).

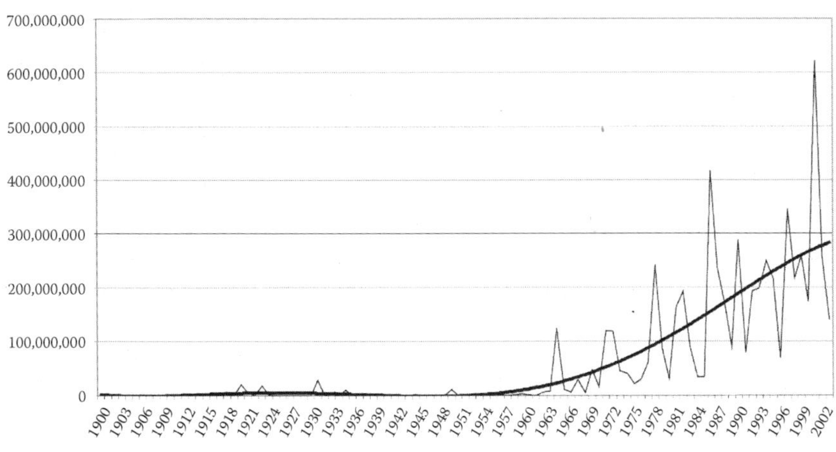

FIGURE 2.1 Total number of people affected by disasters throughout the world (1900 – 2004). (*Source*: EM-DAT: The OFDA/CRED International Disaster Database, www.emdat.be, Université Catholique de Louvain, Brussels, Belgium.)

What is certain about disasters is that they will continue to overwhelm the systems that have been developed to manage them. This is what distinguishes these events as disasters to begin with. With limited resources, labor, and equipment, societies can only plan for events of limited scope. In addition, when those limits are exceeded — when levees fail, warnings come too late, or evacuation is just not possible — the fate of individuals ultimately lies in the actions they have taken to ensure their own survival. Disaster preparedness education equips them with the knowledge they need to prepare most effectively for such circumstances.

This chapter examines basic concepts of disaster management and expands upon those concepts to address specifically the needs of disaster preparedness education campaigns. A brief discussion of risk and vulnerability, and the history of disaster management, is provided for context. An explanation of various government emergency management structures follows. Finally, several relevant terms used throughout this text are defined.

FUNDAMENTAL EMERGENCY MANAGEMENT CONCEPTS

The concepts detailed in this section are provided to explain the basis of the emergency management discipline.

Risk

The term *risk* is simply defined as the chance something bad will happen, and the associated outcome of that possible event. Because everything in life, good or bad, involves risk of some form, risk may be considered unavoidable. Risk affects all people without exception, irrespective of their geographic location or socioeconomic conditions. Individuals and societies alike must constantly make choices that involve known and unknown factors of risk, with absolute risk avoidance generally impossible to achieve. Each of life's decisions — where we live, how we live, what we eat, where we travel, where we work — plays into how much and what kinds of risks we ultimately face.

The concept of risk can have drastically different connotations to different groups, depending upon the context in which it is used. For instance, investors may see risk as an opportunity to make a significant profit. A patient may see the risk associated with an experimental treatment as the opportunity for a life without illness that otherwise could not be cured.

For disaster managers, however, the term's meaning is always negative. However, even among risk managers, there exists no single accepted definition of risk. One of the easiest illustrations of the term, preferred by many emergency managers (and which pertains to its use in this text), is illustrated through the equation stating that risk equals the likelihood of an event occurring multiplied by the consequence of that event, were it to occur (Ansell and Wharton, 1992).

$$\text{Risk} = \text{Likelihood} \times \text{Consequence}$$

In this equation, likelihood is expressed either as a probability (e.g., .15; 50%) or as a frequency (e.g., 1 in 1,000,000; 5 times per year), whichever is appropriate for the analysis being considered. Consequences are a measure of the effect of the hazard on people or property. This equation is useful because it is easy to imagine how the reduction in either the likelihood of a hazard or the potential consequences that might result from it would both result in the reduction of risk. Likewise, any action that increased the likelihood or consequences of a hazard would increase that hazard's risk.

Hazard

All risks are the result of the existence of one or more hazard. The term *hazard*, believed to be drawn from either the French word *hazard* (a game of dice predating craps) or the Arabic *al-zahr* (literally "the die"), is rooted in the concept of chance. Under this interpretation of the word, hazards are events or physical conditions that have the *potential* to cause fatalities, injuries, property damage, infrastructure damage, agricultural loss, damage to the environment, interruption of business, or other negative consequences (Federal Emergency Management Agency [FEMA], 1997; see Figure 2.2). Risk and its associated vulnerability ultimately determine if and when a hazard causes a disaster.

A unique and wide range of hazards termed a "hazard portfolio" influences every individual, family, community, and country. Within this hazard portfolio, there are many methods by which hazards may be categorized. Generally, they are natural, technological, or intentional, although several crossovers exist between these groups. Factors including topography, industrial activities (both within and around the community), behaviors, laws, demographics, climate, income, and many others

FIGURE 2.2 Avalanche chute on Cotopaxi Volcano, Ecuador. This hazard is only a risk if humans place themselves or their property in its path (Coppola, 2003).

determine not only what hazards fall within this portfolio, but also how small or great the risk posed by each will be.

Vulnerability

Vulnerability measures how susceptible an object, area, individual, group, community, country, or other entity is to the consequences of a hazard. Vulnerability, which is closely related to risk but determined by a completely different set of variables, explains how a single event with devastating potential, affecting two individuals or two communities with identical force, presents as a minor issue in one case and a disaster in the other. When assessing hurricane damage, for instance, it is not uncommon to come across a house that has been destroyed immediately adjacent to one that is unscathed. On a larger scale, one can consider how two earthquakes, of almost equal magnitude and intensity, caused fewer than 100 deaths in Los Angeles but more than 20,000 in Gujarat, India.

The etymology of the term *vulnerability* points to the Latin *vulnerabilis*, meaning "to wound." Its measurement is derived from a combination of physical, social, economic, and environmental factors or processes. Vulnerabilities may be decreased through actions that lower the propensity to incur harm, and increased through actions that increase that propensity. For instance, retrofitting a building to withstand the shaking effects of an earthquake will lower that building's vulnerability to the hazard, thereby lowering risk (*resilience*, the opposite of vulnerability, is a measure of propensity to avoid loss). Populations have vulnerabilities as well, which are raised or lowered according to their practices, beliefs, and economic status. Public education is the most effective way to influence the vulnerability of populations and, likewise, individuals.

Disaster

When a hazard risk is realized, individuals and communities are left to manage the consequences. If the event overwhelms the individual's or the community's capacity to respond, the result is a disaster. Not all adverse events are disasters, only those that overwhelm response capacity. Vulnerability and the capacity to manage consequences (dictated by preparedness and mitigation) determine whether a disaster or an emergency event results. The distinction between an **emergency event** and a **disaster** is important. For instance, a simple house fire requires response by a jurisdictional fire department. There is surely property loss, and likely the possibility of injury or loss of life. However, as fires are routine occurrences that are easily managed, they normally are not considered disasters. In the great Chicago fire of 1871, on the other hand, more than 2000 acres of urban land were destroyed over the course of three days. Overall, the destruction included 28 miles of road, 120 miles of sidewalk, 2000 lampposts, and 18,000 buildings, all totaling more than $200 million in property damage (one-third of the value of all property in the city at the time) (Wikipedia, 2005). Between 200 and 300 people died. While both events are fires, only the Chicago fire can be called a disaster.

Disasters also grow in intensity as they overwhelm progressively larger response units. When individuals are unable to manage the consequences of an event, they call upon their local fire department, police department, or emergency medical services (see Figure 2.3). If these agencies are unable to adequately contain the event, they likewise request assistance from the state or provincial government. If that level is unable to manage the event, the disaster becomes national in scope, thereby

FIGURE 2.3 Hotel fire in Bangkok, Thailand. Incidents become disasters only when the capacity to manage their consequences is exceeded (Coppola, 2007).

requiring the intervention of the national government. In situations in which a national government or several national governments are unable to manage the consequences of an adverse event, the event becomes an international disaster, requiring intervention by a range of international response and relief agencies.

Disasters are measured in terms of lives lost, injuries sustained, property damaged or lost, and environmental degradation. These consequences manifest themselves through direct and indirect means, and can be tangible or intangible (see Sidebar 2.1). Disasters may be sudden onset or "creeping." Sudden-onset disasters often happen with little or no warning, and most of their damaging effects are sustained within hours or days. Examples include earthquakes, tsunamis, volcanoes, landslides, tornadoes, and floods. Creeping disasters occur when the ability of response agencies to support people's needs degrades over weeks or months, and they can persist for months or years once discovered. Examples are drought, famine, soil salination, epidemics, and erosion.

Sidebar 2.1: Direct and Indirect Losses

Direct losses, as described by Keith Smith in his book *Environmental Hazards,* are "those first order consequences which occur immediately after an event, such as the deaths and damage caused by the throwing down of buildings in an earthquake" (1992). Examples of direct losses are:

- Fatalities
- Injuries
- Cost of repair or replacement of damaged or destroyed public and private structures (buildings, schools, bridges, roads, etc.)
- Relocation costs/temporary housing
- Loss of business inventory/agriculture
- Loss of income/rental costs
- Community response costs
- Cleanup costs

Indirect losses (also as described by Smith, 1992) may emerge much later and may be less easy to attribute directly to the event. Examples of indirect losses include:

- Loss of income
- Input/output losses of businesses
- Reductions in business/personal spending ("ripple effects")
- Loss of institutional knowledge
- Mental illness
- Bereavement

Tangible losses are those for which a dollar value can be assigned. Generally, only tangible losses are included in the estimation of future events and the reporting of past events. Examples of tangible losses include:

- Cost of building repair/replacement
- Response costs
- Loss of inventory
- Loss of income

Intangible losses are those that cannot be expressed in universally accepted financial terms. This is the primary reason that human fatalities and human injuries are assessed as a separate category from the cost measurement of consequence in disaster management. These losses are almost never included in damage assessments or predictions. Examples of intangible losses include:

- Cultural losses
- Stress
- Mental illness
- Sentimental value
- Environmental losses (aesthetic value)

Safe

One of the most important, but also one of the most contentious, concepts pertaining to emergency preparedness education is the presumption of safety. While the term *safe* might seem obvious as to require no further clarification, its context in regard to emergency management is not evident without a solid understanding and application of risk. Most people wrongly assume that a reference to something as being "safe" implies all risk has been eliminated. However, because such an absolute level of safety is virtually unattainable in the real world, emergency managers and societies establish thresholds of risk that define a frequency of occurrence below which those societies need not worry about the hazard. A practical definition is provided by Derby and Keeney (1981), who contend that a risk becomes "safe," or "acceptable," when it is "associated with the best of the available alternatives, not with the best of the alternatives which we would *hope* to have available" (emphasis added). We are reminded that all aspects of life involve a certain degree of risk. However, as a global society, we are constantly assessing and reassessing what constitutes acceptable risk for every hazard considering that which science, technology, and law can offer to treat those risks.

THE MANAGEMENT OF RISK

For most of life's risks, management is an individual responsibility and is therefore performed on a personal level. Likewise, the consequences of failing to take adequate precaution by preparing for or mitigating these risks are also personal or small in scale (even if *catastrophic* to the individual). Examples of personal risks are easy to imagine, and can be grouped into the following categories:

- Health risks (e.g., eating well, exercising)
- Economic or financial risks (e.g., investments, business ventures, employment decisions)
- Social risks (e.g., public behavior, reputation)
- Physical or safety-related risks (e.g., riding a bike, driving a car, crossing the street, cooking with a gas stove)

On the collective scale, communities, nations, and entire societies face risks posed by a range of hazards whose single events occur on a much larger scale. Earthquakes, floods, volcanoes, terrorist attacks, and pandemics are but a few examples from a long list of possibilities. While annually these large-scale hazards generally result in lower absolute (population-wide) consequences than the consequence sum-total from all individually faced hazards, they are more often given greater attention by governments due to their spectacular nature (wherein individual incidents result in multiple casualties and significant property and environmental damage). Moreover, many of these hazards can be so great in magnitude that when they do occur, they overwhelm not only the individual's capacity to manage them, but also the management capacity of the local or even the regional emergency services — a disaster by definition.

When considering large-scale hazards, vulnerability is most effectively reduced through emergency management efforts conducted at the community, regional, or national government levels. For most of these hazards (though not without exception) it is the government's statutory responsibility to manage (or at least guide the management of) hazard risk reduction measures. When actual disasters result from these hazards, it is likewise a government responsibility to respond to them and provide recovery assistance.

Despite any perceived or actual responsibility that citizens demand of their elected leadership, each individual citizen maintains the power to reduce further his or her own risk levels — even from the greatest of hazards. Research has indicated that within communities where citizens are individually prepared to manage hazard consequences, the community as a whole enjoys much greater levels of resilience (thereby suffering proportionally fewer casualties). On the individual level, people are responsible for managing the risks they face as they see fit. For some risks, management may be obligatory, as is the case with automobile speed limits and seat-belt usage. For other personal risks, such as those associated with many recreational sports, individuals are free to decide the degree to which they will reduce their risk exposure, such as wearing a ski helmet or other protective clothing. This is most apparent in the field of public health, where disease risk clearly affects humans as individuals — prompting a management scheme that promotes management of health risks at the individual level. The public health discipline, for this very reason, is the most experienced with promoting risk management on the personal level. However, by employing the same personal risk reduction philosophy and techniques for all hazards, including the catastrophic ones, individuals will reduce their own and their community's vulnerability to them.

THE MANAGEMENT OF EMERGENCIES AND DISASTERS: EMERGENCY MANAGEMENT FUNCTIONS

Comprehensive disaster management in modern society is based upon four functional components:

1. Mitigation
2. Preparedness
3. Response
4. Recovery

The application of these four functions is often depicted in a cyclical manner, with mitigation and preparedness occurring before a disaster, and response and recovery following it. In practice, however, the actions involved with each function are intermixed with those of the others, and are each performed to varying degrees before, during, and after disasters occur. Disasters are best thought of as occurring along a continuum, with the knowledge and experience gleaned from one disaster influencing preparedness, mitigation, response, and recovery actions taken in the next disaster to occur. Each of these functions is described in the following sections.

Mitigation

Mitigation involves the reduction or elimination of one or both of a hazard's risk components (likelihood and consequence). Sometimes called the "cornerstone of disaster management" (FEMA, 2005), mitigation seeks to "treat" hazards such that they affect society to a lesser degree. While the three other components of the disaster management cycle (preparedness, response, and recovery) are performed either as a reaction to hazards or *in anticipation* of their consequences, mitigation measures seek to reduce the likelihood or consequences of hazard risk before a disaster ever occurs. Of course, mitigation is by no means the simple answer to a hazard problem. In fact, because of the difficulties often associated with most mitigation measures, it is only in the last few decades that their full risk reduction potential has been recognized. In practice, mitigation measures tend to be costly, disruptive, time consuming, and socially unpalatable. They usually carry their own inherent risk and do not always work as intended. Political will for mitigation is hard to come by in many situations, and the public's attention span tends to be too short to accommodate the significant life changes that may be necessary for mitigation to work. However, as the practice of mitigation grows, it will continue to emerge as a means of measurably reducing the incidence of many types of disaster.

Preparedness

Preparedness involves equipping people who may be affected by a disaster, or who may be able to help those affected, with the tools to increase their likelihood of survival and to minimize financial and other losses. This is achieved by training and equipping response agencies at each government level, and by educating the public about what actions they can take to reduce their individual vulnerability and risk. Many different organizations and individuals, including emergency response agencies, government officials, businesses, and citizens, conduct disaster preparedness activities. Each has a unique role to play and unique responsibilities to fulfill when disasters strike. The range of activities that constitute the preparedness function is expansive, and these actions are often the primary factors that determine whether the response is successful. The goals of preparedness are: (1) knowing what to do in a disaster's aftermath; (2) knowing how to do it; and (3) being equipped with the right tools to do it effectively. Preparedness minimizes hazards' adverse effects through effective precautionary measures that ensure a timely, appropriate, and efficient organization and delivery of response and recovery action. Preparedness actions and activities can be divided according to recipient. The government component, which includes administration, emergency management, public health, and other services agencies, is one group. Individuals and businesses are the second group. Preparedness of the first group is normally defined and conducted through the creation and application of an emergency operations plan (EOP) and bolstered by training and exercises. Public education campaigns guide the preparedness of the second group.

Response

Response is characterized by actions taken to reduce or eliminate the impact of disasters that are about to occur, are occurring, or have already occurred (see Sidebar 2.2). Ultimately, the scale of the disaster dictates the response. These actions are aimed at limiting injuries, loss of life, and damage to property and the environment. Response includes not only those activities that directly address immediate needs — such as first aid, search and rescue, and shelter — but also systems developed to coordinate and support such efforts. Furthermore, this function facilitates the rapid resumption of critical infrastructure (such as opening transportation routes, restoring communications and electricity, and ensuring food and clean water distribution) to allow recovery to take place, reduce further injury and loss of life, and speed the return to a normally functioning society. Response processes begin as soon as it becomes apparent that a hazard event is imminent and lasts until the emergency is declared over. Response is by far the most complex of the four functions of emergency management because it is conducted during periods of very high stress, in a highly time-constrained environment, and with limited information. During response, wavering confidence and unnecessary delay directly translate to tragedy and destruction. Relief, the provision of vital necessities to impacted victims, is a major subcomponent of the response function.

Recovery

Recovery involves the repair, reconstruction, or regaining of what has been lost as a result of a disaster and, ideally, a reduction in risk from a similar catastrophe in the future. The recovery phase generally begins after the immediate response has ended, and can persist for months or years thereafter. In a comprehensive emergency management system, which includes pre-disaster planning, mitigation, and preparedness actions, recovery actions may begin as early as during the planning phase, long before a disaster occurs. The actions associated with disaster recovery are the most diverse of all the disaster management functions. The range of individuals, organizations, and groups that are involved is also greater than in any other function (although these participants are much more loosely affiliated than in other disaster management functions). Because of the spectacular nature of disaster events and because disaster consequences affect so many peoples' lives, recovery generates the greatest amount of interest and attention from the world community as a whole. In relation to the other disaster management functions, it is by far the most costly. Disaster recovery is also the least studied and least organized of all of the disaster management functions, and therefore the most haphazardly performed.

Sidebar 2.2: The Disaster Event Lifecycle

The evolution of individual emergency and disaster events can be divided into three phases, each with a corresponding set of preparedness and response activities. These phases are

1. *Prehazard.* During this period of the emergency, the hazard event is impending and may even be inevitable. Recognition of the impending hazard event may or may not exist.
2. *The emergency: Hazard effects ongoing.* This period begins when the first damaging effects begin, and extends until all damaging effects related to the hazard and all secondary hazards cease to exist. It may be measured in seconds for some hazards, such as lightning strikes or earthquakes. However, for others, such as floods, hurricanes, wildfires, or droughts, this phase can extend for hours, days, weeks, or even years. During this time, responders address the needs of people and property as well as the hazard effects.
3. *The emergency: Hazard effects have ceased.* During this final phase of the emergency, the hazard has exerted all of its influence, and negligible further damage is expected. Responders are no longer addressing hazard effects, so their efforts are dedicated to addressing victims' needs, managing the dead, and ensuring the safety of structures and the environment. The emergency still exists and the situation still has the potential to worsen, but the hazards that instigated the emergency are no longer present.

THE MANAGEMENT OF EMERGENCIES AND DISASTERS: EMERGENCY MANAGEMENT STRUCTURES

In most countries, including the United States, emergency management is practiced at all government levels, whether local, regional (state or province), or national. Hazardous *events*, or those situations that require some form of emergency response, occur daily and include house fires, minor floods, or chemical spills, for example. These events are almost exclusively managed at the local level, by the local emergency services, namely fire departments, police departments, emergency medical services, and offices of emergency management. Each of these organizations, and several other peripheral organizations and agencies that participate in emergency management activities, is discussed in the following sections.

The Fire Department

Fire departments (also known as "fire brigades" or "fire services") are the most common local emergency management structure found throughout the world. Although these agencies were first developed in reaction to structural and other forms of fire (the most common hazard faced by communities on a daily basis), many nations' fire departments have expanded their abilities to address a wider range of both regular and rare hazards. Examples of the actions commonly performed by fire departments include:

- Disaster response coordination
- Emergency management
- Emergency medical services
- Fire and arson investigation
- Fire and structural safety inspections
- Fire suppression (structural, brushfire, wildfire, hazardous material fire)
- Hazard and other warning issuance
- Hazard prevention activities
- Hazardous materials response and cleanup
- Issuance and enforcement of permits
- Public education
- Public relations
- Rescue (urban, swift water, wilderness, cave, airborne, alpine, dive, crack and crevice)

FIGURE 2.4 Falls Church, Virginia, Fire Department officials perform an emergency response exercise (Coppola, 2008).

- Terrorism (actual or threatened) response
- Training
- Vehicle extraction

Fire departments may be organized at either the local, regional (county or province, for example), or national level (see Figure 2.4). In the United States, fire departments exist almost exclusively at the local level. How a fire department is structured organizationally often depends on whether the fire service personnel are paid and the kind of government structure in place. Fire departments' needs are driven by both community risk and funding access. Funding sources and levels differ from community to community, but rarely exist at levels sufficient to fully address recognized fire risk and meet all of a fire department's needs, including:

- Personnel
- Training
- Equipment, namely:
    - Apparatus (vehicles specially designed for the firefighters' needs)
    - Firefighting equipment

    - Personal protective equipment (PPE)
    - Rescue equipment
- Communications systems, namely
    - An emergency notification system enabling the public to inform the fire department of an emergency (e.g., 911 system)
    - A radio communications system, which allows responders to talk to each other and to their command center
    - A broadcast system, which allows responders to communicate with the public
- Facilities
- Information
- Authority (statutory)

Law Enforcement

Law enforcement agencies, known as police departments or constabularies, are government-sanctioned entities responsible for maintaining law and order within the community. Though crime fighting is their primary responsibility, law enforcement officers regularly participate in all four emergency management functions. Law enforcement emergency management responsibilities may include:

- Assessment
- Bomb removal and disposal
- Crime fighting
- Crowd control
- Disaster scene security
- Investigations
- Search and rescue
- Security at critical facilities
- Traffic control
- Warning issuance

Although law enforcement agencies are most often centralized and managed at the national government level (e.g., France), many governments (especially federalized systems, such as in the United States) maintain locally based and organized forces. Still others, such as Canada, maintain a mixture of organizational police forces (the Royal Canadian Mounted Police have national jurisdiction, while three provinces and several cities maintain their own local forces). In centralized systems, police officers may be based within their local communities or assigned to communities other than their own. Police forces have traditionally enjoyed more secure and greater funding than other first-response officials, including fire and emergency medical departments.

Emergency Management

Offices of emergency management perform mitigation, preparedness, response and recovery planning, and coordination for large-scale events. This unique field is rather new, appearing primarily during the 1950s civil defense era when governments began preparing for nuclear war. Over time, these offices began to address other catastrophic hazards, and even assumed response and recovery coordination responsibilities. Today, most countries maintain an emergency management capacity at the national level, which addresses mitigation of and preparedness for major disasters. Several countries, including the United States, have empowered their national emergency management offices with response and recovery capacities as well, usually in support of local efforts rather than in place of them. At the local and regional levels, it is common for the fire or police departments to handle emergency management planning and coordination, although many major cities, and all U.S. states, have created dedicated emergency management offices.

Emergency Medical Services

Emergency medical services, often called EMS or the ambulance service, are a specialized form of medical care performed at the scene of a disaster or emergency event. EMS personnel (or emergency medical technicians [EMTs]) are highly trained professionals who offer medical assistance greatly exceeding basic first aid. EMTs stabilize victims for transport to a hospital, where better equipment and conditions are present. Although many police and fire officials are trained to provide first aid and medical assistance, EMS organizations are usually trained and equipped to go beyond the basics, and may even be certified to perform invasive procedures or to administer a range of drugs. EMS systems differ in their level of training, availability of funding, and quality of equipment. In most countries, EMS services are private and charge victims a fee for their services. They may be publicly funded and associated with a hospital or a fire department, or be an independent public service. The vast majority of EMS officials are volunteers.

The Military

Almost every country includes the military in its overall disaster management planning and operations process. Though most democratic governments hesitate to utilize their military resources to address domestic issues, such forces tend to be best suited in many ways to meet the needs required when responding to major disaster. They have secure budgets, specialized equipment, and a trained and quickly deployable workforce, are self-sufficient, and have a highly organized, hierarchical structure. The connection between the military and emergency management goes beyond mere coincidence or convenience for many countries. Emergency management grew out of a defensive need, and the military has been involved throughout the course of that evolutionary process. As such, their status as a valuable resource is widely recognized and often seen as the ultimate last resort. One of the primary concerns of involving the military in disaster response is that of authority. Military forces work with a command structure that can be at odds with the chain of command outlined in most emergency response plans. Additionally, military training optimizes behavior appropriate in hostile, foreign environments, so civilian–military interface during disasters can quickly become contentious if the proper mechanisms and training are not in place to guide such action and prevent conflict. In the United States, laws limit the extent of military involvement in disaster response. However, the state governors have been given direct authority over National Guard resources, whch they call upon regularly to tap into an extensive inventory of heavy equipment and manpower. Utilization of the U.S. Army, Air Force, Navy, Marines, or Coast Guard is not as simple or direct.

Other Emergency Management Resources

There is a full range of non-traditional government agencies and other resources that address one or more of the four emergency management functions. Why, when, and how these agencies participate is determined by the particular characteristics of each community and each disaster. These agencies and offices are described using a full range of titles, although variants within each government structure perform many of the same or similar activities despite differences in nomenclature. These "other resources," which may exist at the local, regional, or national levels, include (but are not limited to):

- Building or Housing Department or Authority
- Department of the Environment
- Department (Ministry) of Public Health
- Department of Agriculture, Forests, Fisheries, and Food
- Department of Civil Defense
- Department of Communications
- Department of Development
- Department of Education
- Department of Energy
- Department of Foreign Affairs, State, or the Exterior
- Department of Labor
- Department of Public Affairs
- Department of Public Safety
- Department of Public Works
- Department of the Interior or Home Affairs
- Office of the Coroner
- Office of the lead government executive
- Transportation Department or Authority

GOVERNMENTAL PREPAREDNESS ACTIONS

There is a diverse range of activities taken by governments to prepare for emergencies and disasters. These actions can be grouped into five general categories, including:

1. **Planning** — In the event of a disaster, each government jurisdictional level will be expected or required to perform a range of tasks and functions in the lead-up to its aftermath. Well in advance of a disaster, governments must know not only what they will need to do, but also how they will do it, what equipment they will use, and how others can and will assist them. The most comprehensive methodology used to plan for disasters is the creation of a community or national EOP. These plans can be scaled up or down depending upon the needs of the community and the particular disaster, and are able to accommodate the complex and diverse needs of a full range of disaster response and recovery actions.
2. **Exercise** — Response exercises allow those involved in emergency and disaster response, as defined in the EOP, to practice their roles

and responsibilities before an actual event occurs. Exercises also help agencies find EOP problems in non-emergency situations, and introduce individuals and agencies involved in response.

3. **Training** — Disaster response officials are more effective when adequately trained to do their jobs. Novice or poorly trained responders add to the possibility of a secondary emergency or disaster, and further strain response resources by diverting officials to manage responder rescue and injury care. Though first responders are likely to have some basic standard of introductory training, the specialized instruction required of disaster response is much more technical and not universally available. Many nations, including the United States, have established centralized or regional training facilities to bring these skills to the local level.
4. **Equipment** — Specialized tools and equipment designed for disaster response and recovery have helped to drastically reduce the number of injuries and deaths and the amount of property damaged or destroyed as result of disaster events. This equipment has also increased the effectiveness of response agencies by protecting the life of the responders themselves. Access to this equipment depends on the availability of funding. Oftentimes, several neighboring communities will share disaster-specific equipment under mutual aid agreements, thereby extending the reach of rarely used, expensive items.
5. **Statutory authority** — In order to ensure that all individuals and agencies involved in emergency management are able to function effectively, proper statutory authorities must exist. Statutory authorities form the legal framework that ensures emergency and disaster response agencies and functions are properly established and staffed, and that they receive regular funding. These laws and regulations also guide how government response officials and agencies may interact with the public and businesses in a range of situations. Response requires travel over and operation on both public and private land, and statutory authorities allow this to occur without question or unnecessary delay. Mutual aid agreements between neighboring communities, states, or even countries also require a legal contract or similar instrument to guide expectations, command, coordination, and reimbursement for services.

INDIVIDUAL AND BUSINESS PREPAREDNESS

In the event of a disaster, it is assumed that governmental resources will be stretched to the limits and even exceed their capacity during the first few hours or days of response. During this time, the public, businesses, and other NGOs must be prepared to provide for their own immediate response needs. At the individual level, public preparedness efforts serve to empower ordinary citizens to help themselves, their families, their neighbors, or complete strangers. Business and other NGO preparedness serves to preserve the integrity and viability of the organization and to protect all human, material, and informational resources.

To be effective, all individuals and organizations involved must do more than simply raise their hazard awareness. A prepared public must also be given the skills that allow them to perform specialized response actions such as search and rescue, first aid, or fire suppression. In recent years, disaster managers have established effective ways to increase public knowledge of disaster preparedness and response activities and got the public to act upon that knowledge. It was long believed that the public was incapable of acting rationally in the face of disaster, but these attitudes have been proved baseless. Studies of actual post-disaster scenarios found that the public acts rationally and effectively, even when frightened or stressed.

JURISDICTIONAL MANAGEMENT AND CONTROL: DEFINING RESPONSIBILITY

Public preparedness education may be carried out by any individual or organization. At the government level, however, such a task may be required or even mandated by local statutes. A community's EOP, or its hazard mitigation plan (HMP) if one exists, may include pre-disaster responsibilities. In most EOPs and HMPs, public education is listed under the responsibilities of a designated official in either the fire department or the office of emergency management. These plans rarely provide any detail about what that task entails, leaving the task for interpretation to the responsible official or agency. With the shortfall of financial and human resources in virtually all emergency services throughout the nation and the world, this function tends to be given low priority or is performed in a cursory fashion. Likewise, many public officials performing this function are without the training or technical knowledge to perform an effective public education campaign.

Success stories of public education campaigns that are funded, developed, and managed by local government agencies do exist, however. Agencies that recognize the value of public education and possess the resources to carry out an effective campaign have experienced varying yet positive success in their endeavors ranging from good to great. Examples of government-funded and -conducted public education campaigns include:

- Washington State "Prepare in a Year" Campaign — Guides personal and family hazards preparedness by promoting a schedule of preparedness tasks that demands no more than one hour per month over a 12-month calendar year (http://www.emd.wa.gov/preparedness/prep_prepare_year.shtml)
- California "Bear Responsibility" Campaign — Promotes a 10-step emergency preparedness program that includes risk identification, making a plan, making a kit, learning skills, mitigation, and more (http://www.oes.ca.gov/Operational/OESHome.nsf/ALL/7A43A30DB8F1F1CA88256FE90079CEB6?OpenDocument
- Alabama "Be Ready" campaign — Includes separate preparedness information and activities for seniors, businesses, families, infants, and children (including an intensive week-long "Be Ready Camp" where children learn several emergency management skills) (http://www.readyalabama.org)

More examples of publicly funded and conducted emergency preparedness campaigns can be found in Chapter 7.

Public education is also conducted largely by the non-governmental and private sectors. Many NGOs consider disaster preparedness issues as central to their mission, or directly related to their individual programs. For instance, NGOs working with special needs populations, such as the elderly, children, or the poor, might consider the higher vulnerability to hazards often faced by these groups — a development issue that their expertise could address. Community foundations, to a growing degree, have begun to fund grant requests for projects that seek to increase disaster resilience of one or more groups within the community or of the community as a whole. In addition, several NGOs that have facilitated the long-term recovery in disaster-affected areas are working to decrease the likelihood of a repeat disaster among the affected population by conducting the education necessary to allow victims to help themselves.

Businesses are similarly becoming involved in public emergency preparedness efforts through a sense of civic responsibility to raise the profile of their brand or their products. Businesses are also educating their own employees in recognition that people victimized by disasters are rarely able to return quickly to work. Most businesses participate in these efforts by collaborating with or providing funding to NGOs or government agencies performing public emergency preparedness campaigns in the communities where the businesses operate or where their sales base is located, but some elect to perform the actual education themselves. Either way, their contribution has become a critical element to the greater goal of public emergency preparedness.

WHAT IS PUBLIC EMERGENCY PREPAREDNESS?

People in almost all nations, and certainly all of those in the industrialized world, have come to expect that their government will intervene in times of disaster to provide life-saving assistance. In response, the elected leadership of these nations have taken great measures, in word and in deed, to assure their constituents that response needs would be met should a disaster occur. Despite these promises, the true test of a nation's response capacity comes only when an actual disaster event occurs. Their success, or lack thereof, is most often a factor of how adequately they have prepared themselves and their public.

Public emergency preparedness equips individuals with the knowledge, skills, or resources necessary to increase their likelihood of survival and to minimize financial and other losses in the event of an emergency or disaster. The justification behind such efforts is the belief that ordinary citizens who are empowered with these tools are better able to help themselves, their families, their neighbors, and their communities. By increasing the resilience of individuals, the collective resilience of the group or population to which they belong also increases.

Individuals take disaster preparedness actions on a daily basis, often without even thinking about them. Glancing at a fire escape diagram upon entering a building, for instance, prepares people with the knowledge required to escape should an alarm sound while they are inside. Scanning the news for weather reports allows people to anticipate any emergency food and water needs. Cutting back overgrown shrubbery around the house provides a firebreak should a wildfire strike. Each of these actions helps individuals to reduce the risk from the unique and wide range of hazards they face.

Public education measures to reduce disaster risk can range from highly specific, addressing a singular disaster consequence, to more general in nature, encompassing the diverse needs of an all-hazards portfolio. For instance, the Institute for Building and Home Safety (IBHS) is well known for its program that assists day-care centers in mitigating the specific consequences associated with seismicity (including falling furniture, books, and fixtures, for instance). On the other end of the spectrum is a FEMA-supported public emergency preparedness initiative called CERT (Community Emergency Response Team) Training. Students enrolled in the CERT course spend several weeks learning the particulars of individual and family planning for emergencies, search and rescue, fire suppression, first aid and disaster counseling, and much more. These skills are designed to prepare students to assist themselves and others in almost any disaster situation they face.

The U.S. Department of Homeland Security (DHS) maintains that public disaster preparedness involves three distinct components:

1. A kit: Making a disaster kit that prepares the individual or family to survive emergencies (where essential resources such as water, food, and clothing are in short supply or not available).
2. A plan: Making a family disaster plan that allows for family members to locate each other, make contact, join together if possible, access disaster information, and make informed decisions.
3. Knowledge: Learning about what hazards exist that affect the individual, how to recognize those hazards when they occur, what the possible consequences of an associated hazard might be, and what actions can be taken to respond to those various disaster scenarios (e.g., evacuation, sheltering in place, moving to a safer location).

This is, in many ways, a bare minimum of preparedness action that is possible, as it fails to address mitigation and recovery planning measures that citizens can take to reduce their hazard risk and facilitate a more effective recovery.

THE ADVANTAGES OF A TRAINED PUBLIC

Public disaster preparedness education, when successful, is a highly effective way to save lives and reduce property damage. In fact, research has found a positive correlation between increased public knowledge about disaster reduction and preparedness basics and a decreased incidence of death and destruction when an incidence of the particular hazard occurs (Foster, 2007). Hazard risk should always be viewed as a dynamic factor (i.e., something that can be altered in a positive or negative way). The action people take will work to either increase or decrease their risk to one or more of the hazards that affect them. By understanding what a hazard is, how to recognize it, and what can be done to mitigate its consequences, individuals can lower their personal hazard vulnerability prior to the occurrence of a disaster. By learning what types of actions can be taken in the event of a disaster, individuals can prepare themselves to respond appropriately to prevent the loss of life and property.

In routine emergency incidents such as car accidents, house fires, and simple criminal acts, local emergency service agencies (fire, police, EMS, and emergency management) are ready and able to assist victims and minimize loss. The magnitude of the consequences associated with major disasters, however, can quickly overwhelm these traditional first-response services, leaving affected residents to fend for themselves for hours or days. By default, these affected individuals become first responders, working to address their own emergency needs and those of their neighbors. Therefore, a prepared public is obviously integral to a community's disaster resilience. When properly informed and educated, members of the general public can learn not only how to recognize a potentially hazardous situation before it occurs, but also about what can be done to minimize risk once that disaster becomes imminent.

A disaster-affected population requires an insurmountable measure of supplies and countless skilled practitioners to address their emergency needs. Lifelines will have been cut, critical infrastructure will be damaged or disabled, and a wide range of injuries and fatalities is likely to have occurred in a matter of minutes or hours. The required response capacity of a community's emergency services is directly proportional to the collective response needs of that community. In other words, the greater the vulnerability of the individual citizenry, the greater the burden of the emergency services in the event of a disaster. Likewise, as individuals reduce their vulnerability, so does the community, resulting in less pressure on its emergency services.

Public preparedness efforts empower ordinary citizens to help themselves, their families, their neighbors, and even complete strangers. To be effective, this effort must go beyond simply raising awareness of a hazard and its affiliated risk. Public disaster preparedness education can decrease individual vulnerability in two primary ways — by teaching individuals how to mitigate their hazard risks, and by training them how to respond effectively when a disaster is imminent or has just occurred.

The public can also be provided with skills that prepare them to perform specialized risk reduction actions prior to and during a major disaster. This includes a wide range of functions — from sandbagging to search and rescue, firefighting to first aid, and many other actions that are described in Chapter 3. Until recently, it was thought that the public was incapable of acting rationally in the face of disaster. Response officials feared that victims would panic or would be unable to use preparedness information effectively. However, studies of actual post-disaster scenarios found that the public could act rationally and effectively, even when frightened or stressed. These studies were what first highlighted the need for governments and other agencies to help the public prepare. During its International Decade of Natural Disaster Reduction, the United Nations (UN) introduced the concept that increased disaster risk awareness among the more vulnerable populations of the world is a vital component of any effective national risk reduction strategy. The UN continues this effort through its International Strategy for Disaster Reduction, which identifies public preparedness as one of four key objectives in establishing greater worldwide disaster resilience.

There have been many situations where an informed public significantly reduced its hazard risk by participating in a public education effort, faring much better than others who did not. Consider the following:

- When the 2004 Boxing Day tsunami events struck Southeast and South Asia, more than 200,000 people were killed within several hours. Many fatalities could have been avoided had precursor warning signs been heeded or had there existed widespread knowledge about how tsunamis formed. There were select communities, however, where previously established risk-communication was credited for minimizing injuries and deaths as compared to neighboring communities that faced similar impacts. For example, the coastal zones of the Indonesian island Simuelue (population 78,000), which sits very close to the source of the tsunami, were inundated by water only eight minutes after the magnitude 9.1 earthquake struck that morning. Many of the island's coastal communities were completely destroyed by the rising water and violent waves, but only seven people died. A Humboldt State University geology professor determined that an oral storytelling tradition, which had preserved preparedness knowledge obtained after a 1907 tsunami, had ultimately equipped the local population with the tools they needed to prevent injury and death quickly and effectively (Cairns, 2005). The affected local population had learned: "Once in a while large earthquakes are followed by large killer waves, so it's always wise to run to high ground and wait a while, just in case." Other populations had much more time to respond, better telecommunications and warning capabilities, and many more resources at their disposal, but fared much worse, experiencing casualty rates as high as 90% in some regions. Survivors interviewed in those places indicated that they had little or no knowledge of tsunamis, were not aware of what caused them, and did not know what typically preceded them.

- In 2002, Ohio's Van Wert County earned the designation "Storm Ready" by meeting several preparedness and mitigation standards established by NOAA. Part of this certification involved the installation of a public alert system and the carrying out of a public education campaign that instructed residents how to react to warnings. On November 10 of that same year, detection systems indicated a high likelihood of tornadoes, and a warning was issued. Several tornadoes materialized, and 35 people were killed across a number of affected states. Even though Van Wert County fell directly in the path of a particularly destructive tornado, there were no deaths and few injuries. In what became a classic example of the value of preparedness education, all of the employees and patrons of a Van Wert County movie theater destroyed by the tornado survived what would otherwise have been a mass casualty disaster. The theater's employees worked with patrons to move everyone into interior hallways and restrooms as local emergency management officials in the Storm Ready campaign had taught them. When the tornado struck just 28 minutes after the warning was issued, the main seating area of the theater (where more than 50 people had just been sitting) was destroyed. Cars were thrown through the theater walls, landing on several rows of seats, and the theater's roof was torn off (NOAA, 2007).
- The CERT program is a U.S. government-supported public disaster preparedness education program that gives detailed mitigation, preparedness, and response training to regular citizens (see

FIGURE 2.5 Charleston County, South Carolina CERT students practice the use of "cribbing" to remove a trapped disaster victim (Jim Tarter, Charleston County Emergency Preparedness Division).

Figure 2.5). CERT, which has grown significantly throughout the country since it was created in 1993, helped to significantly bolster the emergency resources available to several of the towns and cities affected by the 2004 hurricane season (Hurricanes Charley, Frances, Ivan, and Jeanne). CERT-trained citizens were put to use by emergency officials in a number of roles, including staffing emergency operations centers and information line phone banks, registering special needs populations for evacuation assistance, performing evacuations, performing traffic control, distributing ice and water, sandbagging, and conducting damage assessments. CERT teams also helped to staff emergency shelters, clear debris on impassable roads, deliver food, and perform many other activities that would likely have been staff deficient were it not for these pre-trained individuals. In total, more than 4,000 trained CERT team members participated in the response to these events (FL Citizen Corps, 2004).

CONCLUSION

Emergency management is a pivotal function of government. However, the public is a key, if not necessary, stakeholder in the process, with the capacity to do more to save their own lives and those of their friends, families, and neighbors in the first hours of a disaster than any formal emergency services agency. The roots of emergency management and its associated functions and skills are ancient, but without education and training, the public will not be able to incorporate adequately the profession's risk reduction practices. Many governmental, non-governmental, and private entities have attempted to educate the public about emergency preparedness, and have seen great success among their target audiences. Without the correct application of the skills and practices of the communication profession, as detailed in the following chapters, it is doubtful such progress could have occurred.

REFERENCES

Ansell, Jake, and Frank Wharton. 1992. *Risk: Analysis, Assessment, and Management*. Chichester, England: John Wiley & Sons.

Cairns, Ann. 2005. Disaster Lessons: What You Don't Know Can Kill You. Geological Society of America. http://www.eurekalert.org/pub_releases/2005-10/gsoa-dlw101705.php

Derby, Stephen L., and Ralph L. Keeney. 1981. Risk Analysis: Understanding "How Safe Is Safe Enough?" *Risk Analysis*, 1(3), 217–224.

Federal Emergency Management Agency (FEMA). 2005. FEMA Mitigation Directorate. www.fema.gov/about/divisions/mitigation.shtm

Federal Emergency Management Agency (FEMA). 1997. *Multi Hazard Identification and Assessment*. Washington, D.C.: FEMA.

FL Citizen Corps. 2004. Florida CERT — Makes a Difference in the 2004 Hurricane Season. http://www.caloosacert.org/FLCERTMakesADifference122004.pdf

Foster, Chad. 2007. On the line. *Natural Hazards Observer*. 31(3), –11.

NOAA. 2007. "StormReady" Label Proves Correct for Ohio County in Recent Tornado Outbreak. National Weather Service. http://www.weather.gov/stormready/vanwert.htm

Smith, Keith. 1992. *Environmental Hazards: Assessing Risk and Reducing Disaster*. New York: Routledge.

Wikipedia. 2005. "The Great Chicago Fire." http://en.wikipedia.org/wiki/Great_Chicago_Fire

3

The Campaign — Step 1

Early Planning

INTRODUCTION

Emergency preparedness public education campaigns tend to be highly complex endeavors. To be successful, they require adequate planning, staffing, and funding, and must be based upon a comprehensive understanding of the target audience. Far too many efforts have failed as a result of a single major flaw in design: an overlooked cultural taboo, a budget that fell short of actual costs, a lack of community support, or an unreceptive audience.

However daunting these requirements may seem, project failure is rarely inevitable. By subscribing to a regimented planning and project management framework, like the one suggested in this text, communicators can ensure that their message coverage and program outcomes are maximized. Well-laid plans, initiated in the earliest stages of the campaign, allow for the flexibilities often needed when variables and assumptions shift, thereby helping the project team to work efficiently through problems as they arise. The adage that "failing to plan is planning to fail" certainly rings true in the communication field. Moreover, while adequate project planning may seem overwhelming at first, there are always ways to make it happen. The NIH writes that

> Making health communication programs work requires planning, but planning need not be a long-term, time-consuming activity. Nor should all the activities [be conducted all at once], before any other actions are taken. Planning is easiest and best done bit by bit — related to and just in time for the programmatic tasks it governs. For example, you need certain kinds of information about the intended audiences in order to define them, select them, and set objectives. You need different information to guide message development; gather each type as you need it. (National Cancer Institute, 2004)

Proper planning can be the dividing factor between one organization that succeeds in influencing peoples' behavior and another that churns out unread brochures. At the project outset, the planning process may take time, but over the course of the campaign, it will ultimately save time by providing much-needed focus and direction, setting the project on an attainable course. Planning should be considered important for all of the following reasons and more:

- It helps you to better understand the hazards and the associated risk.
- It delineates the role your organization is best suited to play in targeting those hazard risks.
- It helps you to better understand who your audience is, why they are at risk, what they know and do not know about their risk, what they can and cannot do to change their risk, and how they learn about risk.
- It helps you to identify solutions that have a high chance of effecting behavioral change.
- It helps you to determine what you will say, how you will say it, and who will do the talking.
- It helps you to manage your project, prioritize tasks, assign individual roles and responsibilities, and budget the funds required to complete the project.
- It helps you to identify obstacles before they derail your successes.

Chapter 3 through Chapter 6 will present and explain an easy-to-follow structure by which emergency preparedness education campaigns may be planned, conducted, and assessed. Chapter 3 describes the early planning phase, beginning with identification of the problem that is to be addressed — namely the hazard or hazards and their associated risk. This is followed by the critical process of target audience identification, which allows communicators to have a sound understanding of whom they will be communicating with, including their demographics and psychographics (personality and thought characteristics), their vulnerabilities, their perceptions of risk, and the ways in which they receive their information. Using this hazard and audience information, communicators will be able to identify appropriate solutions to be communicated, and to establish project goals and objectives. Early planning will also involve forming a planning team and coalition, identifying and securing partners and stakeholders, identifying potential obstacles, and planning to manage the project.

DEFINE THE PROBLEM

All public education efforts begin with the identification of a problem or problems that are to be addressed. In the case of emergency management, the problem relates to the vulnerability to a known hazard. Each individual and each population group is affected by hazards differently because of many personal and social factors including available cash resources, physical location, knowledge, abilities, perception, age, and gender, among many others. Their behaviors and actions are what define their vulnerability. It is therefore the goal of the risk communicator to determine the source of risk, as well as the actions and behaviors that contribute to vulnerability and what can be done to reduce or eliminate vulnerability.

The process of defining the problem involves the following steps:

1. Identify and analyze the hazard risk.
2. Define the target population.
3. Identify appropriate solutions.

Identify and Analyze the Hazard Risk

The first step in defining the problem is to identify and analyze the hazard that will be managed. Obviously, communicators need to know exactly what hazard they will be addressing if they are to adequately provide the public with risk reduction information. A *hazard* is defined as any event or physical condition that has the *potential* to cause fatalities, injuries, property damage, infrastructure damage, agricultural loss, damage to the environment, interruption of business, or other types of harm or loss. There are many individual hazards that affect us on a personal or family level, such as a fall, minor crime, choking, drowning, or car accident for example. However, disaster preparedness public education efforts focus on those hazards for which the consequences may be so great as to overwhelm the local capacity to respond. These events, by definition, are disasters. Coppola (2006) classifies common hazards according to three categories: (1) natural, (2) technological, and (3) intentional.

Natural hazards are those that originate from natural processes and include the following subcategories:

- Tectonic hazards (earthquakes, tsunamis, volcanoes)
- Mass-movement hazards (landslides, rockslides, debris flows, avalanches, land subsidence, land expansion)
- Hydrologic hazards (floods, drought, desertification)
- Meteorological hazards (hurricanes, tornadoes, ice storms, severe weather, hail, extreme temperatures, windstorms, sand storms, wildfires, thunderstorms, fog)
- Biological hazards (human epidemic or pandemic, animal epidemic, plant epidemic, plagues)
- Other natural hazards that do not fall into any of these categories (meteors, salinization)

Technological hazards are those that are artificial in origin, but for which there is no malicious intent, including:

- Transportation hazards (infrastructure failure or damage, airline accidents, rail accidents, maritime accidents, roadway accidents)
- Infrastructure hazards (power failure, telecommunications failure, computer network failure, water/sewer failure, gas distribution failure, dam failure, food shortage, overburdened public health system, economic failure)
- Industrial hazards (HazMat processing or storage accidents, explosion, fire, raw materials extraction accident)
- Structure fires or failures

Intentional hazards are those for which there exists malicious intent, including:

- Terrorism
- Civil unrest
- Stampedes
- Crime
- War

Generally, if it is the communicators' intention to address the vulnerability to a single hazard — earthquakes, for instance — then there is no need to conduct a full hazard identification for the community. However, if the goal of the campaign is overall emergency preparedness (for all possible hazards), communicators will need to determine all major hazards that threaten the community. This is performed using any of the following information sources:

- Community historical records
- City or county emergency operations plan
- City or county hazard mitigation plan
- Local emergency management officials
- Local libraries
- Local colleges or universities
- Local or county floodplain manager
- Local chapter of the American Red Cross
- Local emergency planning committee (LEPC)

Once identified, a hazard must be profiled and analyzed to determine the likelihood and consequence components of its associated risk, and to provide communicators with a better understanding of how it affects the community. The more completely a hazard is understood, the better that risk communicators can successfully address it among the target population.

A risk assessment tells communicators how often the hazard will likely occur, and what will happen if it does. Risk consists of likelihood and consequence. The risk component can be expressed as a probability (e.g., 50% chance of occurrence in a given year) or a frequency (e.g., occurs once per decade). The consequence component of risk measures the human (deaths and injuries), structural, economic, and environmental effects that would occur if the hazard resulted in a disaster.

Because few organizations have the time, expertise, or resources to conduct a full hazard risk assessment, it is recommended that organizations obtain hazard analysis information from the city, county, or state office of emergency management. Almost all offices of emergency management — which may be a stand-alone department or part of the fire or police departments — will have conducted such an assessment as required by law, and will likely be willing to share all or relevant excerpts of this assessment for the purposes of a public preparedness campaign. These assessments will likely include much of the information that is needed to complete a hazard profile. The hazard profile is a tool that provides communicators with all the hazard information they need to begin identifying a target population and vulnerability reduction solutions in one place, including:

- The hazard name and description
- The hazard frequency or probability, including
    - Historical incidences
    - Predicted future frequency or probability
    - Magnitude and potential intensity
- The geographic range of the hazard (what is affected in the community), including property, infrastructure, and populations
- Duration of the hazard
- Seasonal or other time-based patterns associated with the hazard
- Speed of onset of the hazard
- Availability of warnings

Define the Target Population

A difficult yet pivotal step in planning any communication campaign is the identification and definition of the population to be targeted by its messages. In the hazard identification and analysis step, communicators identify many different populations and groups that are vulnerable to the effects of the hazard in question. However, each group and individual differs with regard to his or her risk and vulnerability; abilities and capacities to mitigate, prepare, respond, or recover; and methods for receiving and processing information. In other words, there can be no single solution that meets everyone's needs equally.

Communicators must limit their message to a single target audience to which the campaign will be tailored, or define each audience individually in order to tailor solutions, messages, and communication methods to each. Because resources are always limited, and it is therefore impossible to reach all audiences effectively, planners must prioritize which audience will be the intended target of their message. In many ways, the particulars of the selected target population determine all other factors — what messages must be developed, how those messages are communicated, what risk reduction options are possible, and what results are likely to be achieved. By learning as much as possible about the target audience, communicators will increase their chances of success immeasurably.

However, limited resources are not the only reason that target audiences are defined beyond the general characteristics of the greater population from which they are found. Target population definition is performed to ensure that message content and delivery are best suited to meet the particular needs of the intended individuals the communicators hope to reach. In the business sector, this process is commonly referred to as *market research*. Market research is performed to define the social, demographic, and psychographic features of a target audience in order to maximize product sales. In public education, the process is almost identical, with the product being an idea or a way of thinking. As is true with any product, the sales pitch behind that idea must be designed for the recipient population — their comprehension, attitudes, biases, and purchasing power, among many other factors.

All audiences, small or large, can be subdivided into smaller and smaller definable groups. The more defined a target audience becomes, the more targeted an education campaign can be tailored to suit those individual characteristics. However, as campaigns are adapted to suit more and more specific audience characteristics, the more involved and more expensive they generally become. It is, after all, up to the discretion of the communicators planning the education effort to decide how particular they wish to make their audience profile.

Audience profiling is so critical because the ways in which people receive messages and interpret information are by no means uniform. For example, while children learn primarily through teachers, peers, and the other members of their families, their parents may be more likely to educate themselves through newspapers, social networks, and television. However, even as a group, parents may differ in how they learn according to their level of education, age, and social network. It is easy to see how each group can differ so much in its learning mechanisms when one considers all of the possible demographic characteristics that define individual members, including socioeconomic status, education, race, gender, age, and employment.

Audiences are categorized primarily by demographics and defining characteristics. The most basic determination is geography — residents of a community, city, county, floodplain, or state, for example. More often, geography is a secondary defining characteristic from within which a smaller segment of the population is determined. The actual group that is chosen will more likely be determined by the interests of the organization conducting the campaign, a particularly high vulnerability among only certain members of the general population, or some other factor that is of special importance to the organization or group conducting the campaign.

The following list is provided to give a sampling of the more common headings under which the myriad identifiable characteristics describing target populations are found. Ultimately, the chosen target population members will be described using defining characteristics from several of these categories. For example, the selected target population could be "poor families with young children in rural areas of Montana."

- **Population by location**
  - Jurisdiction (generally set by political boundaries — either town, city, county, state, or nationwide)
  - Risk area (selection may be based on the location of a community in a highly vulnerable and hazard-prone zone, its accessibility in the event of a disaster, its disaster history, and its local resources)
  - By place of residence or employment
- **Physical or mental ability**: There are approximately 54 million people in the United States classified as having a disability — nearly one-fifth of the population. However, according to a 2003 survey commissioned by the National Organization on Disability, only 42% of emergency managers for states and cities across the country said that they had a public awareness campaign focused on providing emergency information to people with disabilities. Within the category of disabilities, there are defining characteristics that will guide how messages are formed. For instance, people who are deaf have one set of individual preparedness concerns, and they will receive information one way, while those with a physical handicap such as the loss of one or more limbs will have different concerns and will receive their information (as a group) in a different manner.
- **Urban or rural livelihoods**: The type of communal settlement within which a population resides determines not only its risks, but also the means by which its members receive information and what options they have for preparing for disasters. For instance, while residents of cities may benefit from robust media markets, they also face many unique hazard risks not faced by most rural populations (e.g., the presence of critical infrastructure, an increased risk of international terrorism). Rural populations, however, may face risks and conditions unique to their living situation, including fewer emergency management resources and more dispersed social networks. Research has found that natural disasters are more likely to affect non-metropolitan areas, which make up about 75% of the nation's total land area.
- **Income**: Low-income population groups demand special attention because they often lack access to disaster information

through traditional delivery systems like television and radio. Furthermore, low-income groups are more likely to be located in high-risk areas or near infrastructure susceptible to accidents or intentional acts, such as train stations, railroad tracks, and chemical plants.

- **Transience:** Populations that exist within a geographic area who do not consider that area their permanent residence often have special vulnerabilities that differ greatly from that of the general population. Transient populations include tourists, business travelers, student residents, and seasonal workers, for example.
- **Religion:** Members of different religious affiliations will perceive, prepare for, and respond to hazards and disasters in different ways as determined by their religious beliefs and rules. These affiliations will also define how they receive and process information, both important factors to communicators who are planning a public education campaign.
- **Age:** Members of different age groups differ greatly from one another in terms of abilities, perceptions, and many other factors. The most common delineations of populations by age include children, adolescents, adults, and seniors, although there are many other groupings that are possible in order to best meet the needs of communicators (see Figure 3.1).
- **Gender:** While differences between genders are decreasing in the United States with regard to emergency preparedness, there are still differences in the ways that men and women receive and process information, as well as how they perceive risks.
- **Literacy:** Communicators must have an understanding of the education levels of target audience members, as this characteristic will help them to design their message content and means for delivery.
- **Ethnicity:** Ethnicity can be an important factor in determining message communicators, as members of ethnic groups may hold an inherent trust or distrust in certain communicator types (government officials or police officers, for example).
- **Employment or school status:** This characteristic is important because it can help communicators better understand target population members' financial or time resources, and help them determine the best means to reach them. Examples of subgroups

FIGURE 3.1 Sparky the firedog, the official mascot of the National Fire Protective Association, has been an effective tool for teaching fire prevention messages to children (Coppola, 2008).

under this heading include the employed, the unemployed, retirees, workers in specific occupations, and students.

- **Psychographics:** People differ greatly in terms of their psychological profiles. This can include their attitudes, outlooks, self-image, opinions, beliefs, values, and many other personality traits.
- **Health:** People who enjoy good health can differ from those with general or specific ailments with regard to their perception of risk, their ability to change behavior, or other factors.
- **Language:** Language is one of the most fundamental requirements of communication. Communicators must have a full understanding of the language abilities of their target population in order to determine what language is most appropriate for campaign messages.

Other defining characteristics include housing type, family status, preferred method of transportation, business affiliation, culture, and behavior. The number of defining characteristics is almost limitless. While it is important to define the target audience as fully as possible, it is also important not to define it so narrowly that there are few individuals who fit the defined profile.

The most comprehensive source of data and information available to drive the process by which audiences are defined is the U.S. Census. Conducted every 10 years, and estimated periodically between official counts, the official census can provide a wide range of information on populations within given jurisdictional and administrative boundaries. These data are especially useful in determining income, age, gender, household, employment, ethnicity, and many other demographic factors. However, the census will not give communicators psychographic information that helps them understand audience perceptions, attitudes, learning patterns, and other important factors. To acquire this kind of information, communicators will need to rely upon other methods of information collection. Effective examples of information collection include:

- Interviews with members of the target population
- Group meetings with representatives from the target population
- Meetings with or assistance from individuals, organizations, or agencies that regularly serve or work with the target population (e.g., teachers, NGOs, religious organizations, service organizations, corporations, foundations, fire departments, counselors, newspapers)
- Social science research publications
- Data maintained by polling companies or depositories of polling information
- Chambers of commerce
- Advertising agencies, newspapers, and radio and television stations (for media-use data and buying and consumption patterns)

Once the primary audience has been identified, communicators begin to profile the members of this group in order to learn as much as they can about their particular needs, preferences, and characteristics. The more that is known about the primary audience, the better the message can be designed, delivered, and timed. To ensure maximum impact, planners should try to establish a baseline by answering the following questions:

- What people make up this group? Are they children, adults, senior citizens, students, homemakers, business executives, blue-collar workers, single, married?
- What are the special characteristics and needs of this group? Consider such factors as age, education and literacy levels, gender, occupation, motivations, cultural and social interests, activities, and preferred entertainment options.
- What specific hazard consequences (from the hazard in question) affect members of this group?
- How are members of this group at particular risk from the consequences of the hazard (vulnerability)?
- Risk perception (see Sidebar 3.1):
    - What does this group already know about the hazard?
    - What disaster knowledge do they lack?
    - What misinformation (inaccurate beliefs) do they possess?
    - How does this hazard measure up in comparison to the concerns and fears they hold for all hazards?
- What specific characteristics of this audience place them at increased risk from the hazard (see Figure 3.2)?
- What abilities does this group have to address risk and vulnerability?
- What desires does this group have to reduce their vulnerability?
- What social, cultural, or economic obstacles does this group face in minimizing vulnerability?
- What social or cultural factors would help to affect change or influence message delivery?
- What benefits do members of this group associate with behavior change?
- From what sources does this group typically receive information? This could include newspapers, television, radio, mail, town meetings, informal social networks, parents, and peers.
- When are members of this group most receptive to messages?
- Where are members of this group most apt to receive information?
- Who are the most influential voices for this group? Role models, teachers, parents, relatives, leaders. Are there other people in this community to whom this group listens and respects (e.g., elders or clergy), often called *gatekeepers*?

FIGURE 3.2 Tsunami evacuation signs on Kamala Beach, Thailand. The 2004 Boxing Day tsunami killed hundreds of tourists in Thailand who had little knowledge of what to do when a tsunami is imminent (Coppola, 2008).

Sidebar 3.1: Risk Perception

The branch of science that studies why people fear the things they do (and why they do not fear other things) is called *risk perception*. Understanding trends in public risk perception helps to explain why, for instance, millions of people in the Washington, D.C., metropolitan area were so disproportionately afraid of the Washington, D.C., sniper in 2002 even though they were statistically less vulnerable to that than, for instance, automobile accidents, food poisoning, heart disease, or cancer. Risk perception is a primary factor, though not the only factor, that determines whether people prepare for the hazards they face.

In their article "Rating the Risks," Slovic, Fischhoff, and Lichtenstein (1979) begin, "People respond to the hazards they perceive." These scientists discovered that people tend to misjudge their risk according to four risk perception fallibility conclusions, namely:

1. Cognitive limitations, coupled with the anxieties generated by facing life as a gamble, cause uncertainty to be denied, risks to be distorted, and statements of fact to be believed with unwarranted confidence.
2. Perceived risk is influenced (and sometimes biased) by the imaginability and memorability of the hazard. Therefore, people may not have valid perceptions even for familiar risks.
3. Risk management experts' risk perceptions correspond closely to statistical frequencies of death. Laypeople's risk perceptions are based in part on frequencies of death, but there are some striking discrepancies. It appears that for laypeople, the concept of risk includes qualitative aspects such as dread and the likelihood of a mishap being fatal. Laypeople's risk perceptions are also affected by catastrophic potential.
4. Disagreements about risk should not be expected to evaporate in the presence of evidence. Definitive evidence, particularly about rare hazards, is difficult to obtain. Weaker information is likely to be interpreted in a way that reinforces existing beliefs. (Slovic, Fischhoff, and Lichtenstein, 1979)

People tend to fear a hazard risk less as they become better informed with more specific details of the risk. However, the amount a person can discover about a risk will almost never be complete, as the actual likelihood or consequence most risks pose cannot be quantified in a way that addresses the specific threat faced by individuals (even well-known risks such as cancer or heart disease) (Ropeik, 2002). The more uncertainty a risk poses, or as Slovic, Fischhoff, and Lichtenstein state, "the more of a gamble something is," the more people will fear it. In the face of uncertainty, people will consciously or subconsciously make personal judgments based upon imperfect information in order to establish some individual concept of the risk they face (Slovic, Fischhoff, and Lichtenstein, 1979). These judgments, based upon uncertainties and imperfect information, often cause people to wrongly perceive their own risk, more often in a way that *overstates* reality.

People are more afraid of those things that they can imagine or that they can remember. These easily *available* risks, as they are called, tend to be overestimated regarding their likelihood of occurrence. For instance, we rarely hear about a person dying from a common cause such as a heart attack, unless somebody close to us dies of that specific cause. However, the media will often heavily report on a death that is a result of an uncommon cause, like the West Nile Virus. The result tends to be that people underestimate common risks and overestimate rare risks. Generally, people tend to fear what they hear about repetitively or constantly. This phenomenon is referred to as the *availability heuristic*, which states that people perceive an event to be likely or frequent if instances of the event are *easy to imagine or recall*. This is a perception bias that can be correct when considering events that are, in fact, frequently observed, such as in the case of those who believe that automobile accidents are common because almost everyone they know has been involved in one. However, when a risk that is spectacular but not necessarily common receives constant media attention, such as high school shootings did in the 1990s (in particular, the Columbine attack), people often wrongly assume that similar events are very likely to occur.

It can be difficult for people to understand statistics they are given, and even more difficult for them to conceptualize how those statistics apply to them personally. Furthermore, these statistics tend to do little to affect the way people perceive the risks that are calculated. This is not to say that the average person lacks sufficient intelligence to process numbers, it is just that the numbers are not the sole source of influence on public risk perception. It has been discovered through extensive research that people use other, more heavily weighted, *qualitative* factors in addition to the quantitative likelihood of a hazard resulting in personal consequence when ranking their risks (Slovic, Fischhoff, and Lichtenstein, 1979). People are generally more concerned with the consequence component of risk than they are about the likelihood component.

Slovic, Fischhoff, and Lichtenstein, in their article "Facts and Fears: Understanding Perceived Risk," proposed that there are 18 risk characteristics that influence public risk perception. Of these characteristics, 17 fall under two subgroups called *factors*: factors related to dread (Factor 1) and factors related to how much is known about the risk (Factor 2). Using these 17 characteristics, they examined public perceptions of 90 risks and plotted their findings on a two-dimensional graph depicting Factor 1 on the x-axis and Factor 2 on the y-axis.

Factor 1

- Dreaded vs. not dreaded
- Uncontrollable vs. controllable
- Global catastrophic vs. not global catastrophic
- Consequences fatal vs. consequences not fatal

Not equitable vs. equitable
Catastrophic vs. individual
High risk to future generations vs. low risk to future generations
Not easily reduced vs. easily reduced
Risk increasing vs. risk decreasing
Involuntary vs. voluntary
Affects me vs. does not affect me
Not preventable vs. preventable

Factor 2

Not observable vs. observable
Unknown to those exposed vs. known to those exposed
Effect delayed vs. effect immediate
New risk vs. old risk
Risks unknown to science vs. risks known to science

Risks that exhibited the left of each pair of characteristics listed above were seen as more fearful than those that exhibited the right form. For example, uncontrollable risks are more feared than controllable ones.

Slovic, Fischhoff, and Lichtenstein (1979) state that "people's beliefs change slowly and are extraordinarily persistent in the face of contrary evidence. New evidence appears reliable and informative if it is consistent with one's initial belief; contrary evidence is dismissed as unreliable, erroneous, or unrepresentative." They add that "convincing people that the catastrophe they fear is extremely unlikely is difficult under the best conditions." This stoicism is compounded by the fact that once people make their initial judgments, they believe with overwhelming confidence that their beliefs are correct. This phenomenon, called the 'overconfidence heuristic,' states that people often are unaware of how little they know about a risk, and how much more information they need to make an informed decision. More often than not, people believe that they know much more about risks than they actually do.

Risk perception factors into what is called a "worldview." Worldviews are conceptualized as "...general societal, cultural, and political attitudes that appear to have an influence over people's judgments about complex issues" (Slovic, 1999). Studies have found strong correlations between worldviews and risk perceptions (e.g., Dake, 1992; Jenkins-Smith, 1993). Some specific worldviews that have been investigated (e.g., Buss, Craik, and Dake, 1986; Dake, 1991; Jasper, 1990) are

- **Fatalism** — characterized by those who feel that they have little control over their own fate.
- **Hierarchy** — typifies those who prefer to leave risk decisions to the experts.
- **Individualism** — associated with those who believe that those with greater ability should earn more.
- **Egalitarianism** — describes those who feel that the source of many of the world's problems is inequality.
- **Technological enthusiasm** — depicts those who trust in advances in technology to improve health and societal well-being.

While these categories are not mutually exclusive and this list is certainly not exhaustive, the purpose of their inclusion in this chapter is to emphasize the importance of doing formative research about the characteristics of one's target audience that shape their perceptions of risk before trying to craft messages to change their risk-related behaviors. An in-depth discussion of all findings associated with differential risk perceptions is outside the realm of this book, but practitioners can greatly benefit from conducting their own literature searches involving their specific population of interest before designing campaigns. Several theories of communication and psychology provide lists of important elements to consider when engaging in this type of formative research. See Chapter 4 for further discussion.

In relatively recent years, researchers have called attention to the limitations posed by assessing risk strictly in terms of quantifiable scientific certainties. Risk communication scholars emphasize the importance of considering risk perceptions, intuitive risk judgments that citizens rely on to evaluate hazards, in addition to the technologically advanced methods used by analysts when making risk assessments. As argued by renowned risk communication scholar Paul Slovic (1987):

> Lay people sometimes lack certain information about hazards. However, their basic conceptualization about risk is much richer than that of experts and reflects legitimate concerns that are typically omitted from expert risk assessments. Efforts are destined to fail unless they are structured as a two-way process. Each side, expert and public, has something valid to contribute. Each side must respect the insights and intelligence of the other. (p. 285)

Of the variety of audience characteristics that are important to planners, there are some that may facilitate change, others that may hamper it, and yet others that, while they seem important, may have little overall impact. The ability of the planning team to analyze these audience characteristics, and to identify the strengths, weaknesses, opportunities, and obstacles regarding how that audience receives and retains information, is key to the campaign's outcome. The following examples illustrating how particular audience characteristics can influence public education is adapted from Bernstein, Clark-Stewart, Roy, Srull, and Wickens (1994, pp. 627 and 631):

1. **Age**: Younger people are more likely to change their attitudes than older ones, perhaps because they are more receptive to opinion and input from others, and have yet to build a base of experiences that may firm up their own attitudes. The receptivity of children to developing positive attitudes toward emergency preparedness can have a substantial influence on overall community disaster education efforts. An application of this principle can be found in emergency preparedness educational efforts in schools, which can have a powerful effect on how children approach disaster hazards at home.
2. **Intelligence**: Some argue that highly intelligent audiences will understand the persuasive arguments, and thus be more likely to change their attitudes and behavior. Others suggest that such individuals will challenge the logic of the arguments (i.e., "counter-argue") and will be more likely to find flaws in the presentation, and therefore not change at all. Research suggests that the degree to which people focus on understanding and enacting the recommendations rather than counter-arguing against the recommendations is a function of the extremity of the position taken in the argument and how involved audience members are with the topic. Audience intelligence alone is not a reliable correlate with susceptibility to persuasion. However, if the recommendation is not promoting a severe change and if the audience does not seem to hold a firm position on the topic, people of higher intelligence are probably more likely to successfully follow the plan.
3. **Self-esteem**: Those with low self-esteem tend to value the attitudes of others more, but at the same time may be incompletely attentive to events around them. Thus, while susceptible to persuasive

arguments, those with low self-esteem may not think about them enough to effect change. In contrast, those with high self-esteem pay attention to others, but their self-confidence precludes susceptibility to change. The American Red Cross recommends personalizing the disaster preparedness issue by reinforcing for people that they can effectively prepare for a disaster. "Tell them: 'You can do this; you can get ready; it's something you can do now.' People get more involved when they feel they are in control over their situation" (American Red Cross, 1992, p. 80). This technique can also increase self-efficacy (discussed in Chapter 4).

4. **Relevance of the topic/involvement**: If the topic being discussed is highly relevant to the audience, they are more likely to pay attention to the message. Strongly presented arguments are thought to increase source credibility, thereby increasing the effectiveness of the message. Therefore, in most cases, the more important the topic is to those in the audience, the more they will be receptive to strongly presented arguments. However, the increased attention that comes with high involvement is also associated with greater scrutinizing of the message. Therefore, weak arguments are more likely to be rejected by highly involved audience members than by those to whom the topic is not relevant. In contrast, as noted in Chapter 1, those who are not highly involved with the issue are more likely to be persuaded by simple heuristic cues such as the number of arguments presented (regardless of their strength) and the credibility of the speaker.

Communicators often seek to address two very different audience types when they set out to affect a behavior change in a population. In both groups, communicators are trying to bring about a certain behavior, but the type of behavior sought is what places individuals in one group or the other. The first and most obvious group is called the *primary intended audience*. This audience includes those individuals whose hazard vulnerability communicators are trying to reduce. There may be one primary intended audience, or several different primary intended audiences, depending on how much segmentation is performed by planners (see Sidebar 3.2). *Secondary intended audiences*, or gateway audiences as they are also called, are those with influence on the primary intended audiences or those who must do something to help cause the change in the primary intended audiences. This group also requires behavior change, but the type of behavior relates to their interaction with the primary intended audience, not with their hazard vulnerability. For this reason, it is often the case that much different kinds of messages and tools need to be developed if communicators decide to utilize the assistance of this valuable (and sometimes vital) resource.

Sidebar 3.2: Segmentation

Defining subgroups of a population according to common characteristics is called *segmentation*. Segmentation can help you develop messages, materials, and activities that are relevant to the intended audience's current behavior and specific needs, preferences, beliefs, cultural attitudes, knowledge, and reading habits. It also helps you identify the best channels for reaching each group, because populations also differ in factors such as access to information, the information sources they find reliable, and how they prefer to learn. Increase your program's effectiveness by developing strategies that are attuned to the needs and wants of different intended audience segments. In fact, given the diversity of the public, trying to reach everyone with one message or strategy may result in an approach that does not effectively reach those most able or ready to change. Be aware, though, that moving from a mass-market strategy to a differentiated strategy will add economic and staff resource costs for each additional segment. The key to success is to segment the intended population on characteristics relevant to the disaster preparedness behavior to be changed. A logical starting point is the behavior itself. When possible, compare those who engage in the desired behavior with those who do not and identify the determinants of their behavior. Many planners simply rely on demographic, physical, or cultural segmentations. However, people who share these characteristics can be very different in terms of preparedness behavior. For example, consider two 50-year-old African-American men. They are both employed in comparable jobs in the same company, have the same amount of schooling, and almost identical household incomes. Their houses are adjacent to each other, their families are members of the same church, and their families often spend time together socially. Their television and literary interests are very similar, and they subscribe to the same national and local newspapers. Neither has been a victim of a major disaster. Yet one man has worked with his family to create a family emergency plan, has ensured that the family has at least one week of emergency food and water, and has purchased a National Weather Radio. The other, however, has taken none of these actions. A demographic, physical, or cultural segmentation would group these two men together, yet one is a member of the intended audience for emergency preparedness communications about family preparedness, and the other is not.

Identify Appropriate Solutions

Once the hazards have been identified and the target population has been defined, communicators need to begin formulating the solution they intend to communicate. This solution will be a preparedness measure, a mitigation measure, or a combination of both. Moreover, while there are several possible solutions to each hazard vulnerability, the chosen solution will be the one that, given the particular characteristics of the target audience, is most likely to succeed.

Defining solutions understandably requires a working knowledge of both the hazard and the population vulnerable to it, as was determined in the previous two steps. Communicators begin to identify the most appropriate solution by identifying all possible solutions. From this list, they weigh the benefits, costs, and likelihood of audience members taking the proposed actions, in order to select the best alternative.

Mitigation solutions work by decreasing either the likelihood of a disaster occurring or the consequences of a disaster should one actually occur. Preparedness measures allow an individual or group to respond more effectively to a disaster once it happens, through action or equipment. In selecting the appropriate solution, communicators will need to fully understand not only how the mitigation or preparedness option works (from conception to implementation, including maintenance), but how these factors are influenced by the particular characteristics of the target population.

Generally, each mitigation or preparedness option is analyzed according to the following factors, as each relates specifically to the target population:

- Benefit (the amount of actual vulnerability reduction)
- Cost (in financial terms, to the individual)
- Time (required to implement or maintain the solution)
- Availability (of materials, resources, and expertise that are required to implement the solution)
- Secondary negative consequences
- Sustainability

- Target audience obstacles (problems — ideological, cultural, technical, or other — that the population will have with the solution)
- Feasibility obstacles (problems that are independent of the target population that would make implementation of the solution difficult or impossible)
- Likelihood that individual members of the population will take the mitigation or preparedness action
- What, if any, segment of the population is already taking this action, and their successes and failures in doing so

Based upon this assessment, communicators are able to make a more informed decision about what actions will have the greatest overall effect in reducing population vulnerability. Remember that there is no perfect solution, so the option that brings about the greatest change is preferred above the rest.

With the hazard identified, the target population defined, and the most desirable solution singled out, communicators will have successfully defined the problem to be addressed. The entire program will be built upon this foundation. It is not difficult to understand how a lack of knowledge in any of these three areas could make for potentially devastating setbacks later on.

MARKET RESEARCH

At each step in the campaign process, from planning to assessment, there will become a need for communicators to gain more insight and to test the validity of their assumptions and proposed methods. The most effective way to do this is to work directly with a sample group from the target population itself. This is often referred to as *market* or *communication research.*

Market research provides communicators with a much deeper understanding of how the issues with which they are dealing apply to the target audience in particular. When communicators make decisions regarding hazards, solutions, communication methods, and other aspects of the campaign, they are making assumptions about how these issues apply to or affect the target audience. For instance, communicators may need to find out how members of the population feel about making the behavior change they have chosen. If the communicators assumed that the population's members would be receptive to the idea, but through market research discover they are vehemently opposed to it, they will have saved themselves considerable time and money by having the option to change course at this early juncture. By failing to conduct such testing, they may not find out about such attitudes until the campaign has begun and resources have already been dedicated.

By working directly with members of the audience, assumptions are validated or refuted, thereby providing communicators with more realistic impressions of what needs to be done, how successful their efforts' methods or materials will be in practice, or how successful their conducted efforts have been in affecting change. At this early point in the process, communicators use market research methods to learn more about their proposed solution. In order to take the next steps — namely, developing a message and choosing communication methods — it is key to understand as much about the knowledge, attitudes and feelings, misperceptions, and assumptions that the audience holds with regard to the proposed mitigation or preparedness solution. It is within these bounds that the communication campaign will be designed, taking advantage of these factors in planning rather than encountering them unexpectedly along the way.

There are several ways in which market research can be conducted, with the chosen methods a factor of capacity, time, and available funding. Questionnaires are the most common, but other highly effective methods include the following:

- **Focus groups:** A qualitative research technique in which an experienced moderator guides approximately 8 to 10 participants through a discussion of a selected topics, allowing them to talk freely and spontaneously. Focus groups are often used to identify previously unknown issues or concerns, or to explore reactions to potential actions, benefits, or concepts during the planning and development stages.
- **In-depth interviews:** A type of qualitative research in which a trained interviewer guides an individual through a discussion of a selected topic, allowing the person to talk freely and spontaneously. This technique is often used to identify previously unknown issues or concerns, or to explore reactions to potential actions, benefits, or concepts during the planning and development stages.
- **Theater-style testing:** Individuals typical of the intended audience are invited to a conveniently located meeting room. The room should be set up for screening a television program. Participants are generally not told the real purpose of the session, only that their reactions to a television program are being sought. At the session, participants watch a television program. The program can be any

entertaining video approximately 15 to 30 minutes in length. The videotape is interrupted about halfway through by a sequence of four commercials. The emergency preparedness message should be inserted between the second and third commercials. At the end of the program, participants receive a questionnaire and answer questions designed to gauge their reactions, first to the program and then to the advertisements. Finally, the ad is played again and participants complete several questions about it. The majority of these questions should be closed-ended to enable an easy and accurate summary of participant responses.

EXISTING PROGRAM RESEARCH AND GAP ANALYSIS

Before going about planning the campaign from the ground up, it is always wise to identify and assess what has already been done to address the issue, what is currently being done, and what the outcomes of these actions are (in terms of vulnerability reduction and behavior change). There is no sense in repeating the work of others. Nor is it wise to conduct a campaign whose message differs from or even contradicts the message of another campaign without first planning how you will explain the differences. In many cases, even when no other organization or agency has addressed the problem in exactly the same way, it has tackled some part of the problem or addressed similar issues with the same population. Existing program research is conducted for the following reasons:

- To avoid reinventing the wheel
- To build upon the successes of other programs and benefit from the trust they have gained
- To find collaborative opportunities
- To understand and learn from the failures of other programs
- To understand any misconceptions, mistrust, or other incorrect or negative feelings that may exist because of a prior communication attempt

When program research is conducted to determine the individual communication needs or actions lacking within the greater spectrum of communicaton efforts currently underway, it is called a *gap analysis*. Gap analysis looks for specific areas where messages are not reaching audiences. This could be a factor of segments of the population, or it could relate to components of the message received by all members of the population.

DETERMINE PROJECT FEASIBILITY

In Chapter 1, the many possible components of a comprehensive public disaster preparedness campaign were described, including communication, facilitation, funding and financial incentives, policy change, and technology. With a full comprehension of the problem in hand, and a better understanding of how the target population is affected, communicators can better determine which of these components will be necessary to actually bring about measurable vulnerability reduction. If communication alone will do little to change attitudes in the absence of a change in policy, and no effort to bring about a policy change exists within the campaign strategy, it would be better to change the campaign goals to something more achievable than trying to go ahead with the campaign as is. On the other hand, if the target population lacks only the funding and knowledge to bring about change, and a partner organization or sponsor is willing to assist by helping those unable to afford the necessary measures (such as purchasing a weather radio, go-kit, or smoke detector), then this combination would likely bring about much greater change than communication alone.

Once you have determined that your proposed solution has the potential to bring about actual change, there are a few final factors that must be considered before setting out on the full campaign effort. These include the following:

- Does your organization have the necessary expertise and resources to conduct the campaign? If not, can these be acquired?
- Does your organization have the necessary authority or mandate to bring about the changes or measures being proposed?
- How much time does your organization have to dedicate to campaign planning and implementation?
- What, if anything, can be accomplished in that time?

ESTABLISH REALISTIC GOALS AND OBJECTIVES

With the wealth of information communicators have gained in these initial steps, they will finally be able to begin establishing a campaign goal and objectives. The goal and objectives will be used to guide the campaign design and methodology as it progresses. In fact, the purpose of the initial research (the problems, the audience, and the solutions) was primarily to establish exactly what communicators are setting out to achieve (the goal) and how they plan to go about doing it (the objectives).

The campaign *goal* is defined as the general emergency preparedness outcome that the communication team hopes to create. An example of a campaign goal is

> To encourage college freshmen at the University of California to plan and prepare for earthquakes.

All aspects of the campaign will be designed to meet this central goal. The goal does not indicate how, to what level of success, or in what time frame this outcome will be achieved.

Campaign *objectives*, on the other hand, are specific, plainly measurable action points that the communication team hopes to achieve in its drive to meet the goal. They are more specific than goals in that they offer some quantifiable target outcome involving specific knowledge, beliefs, attitudes, or behaviors of intended audience members. An example of a campaign objective that relates to the preceding goal could include:

> To increase by 10% the number of UCLA students who have secured furniture to walls, using anchors, by the end of the fall 2010 semester.

Note that while the use of furniture anchors by *every* student at the university would be optimal, the goal of the campaign is only to *increase* the number by 10%. It is unrealistic to expect that any single campaign could completely solve a problem. The NIH suggests that practitioners seek the guidance of statisticians or emergency preparedness experts to help to determine realistic rates of change before setting quantifiable communication objectives. The organization points out that even commercial marketers consider a 2–3% increase in sales to be a great success.

Objectives should always be attainable, measurable, and time specific. Objectives do not necessarily have to be so closely linked to achievement of the goal, as long as their action ultimately leads to it. For instance, another objective in this same campaign could be

> Conduct one earthquake preparedness presentation, lead by a UCFD firefighter, in each of the freshman campus dorms by the end of student orientation week.

If a single target audience has been selected, then objectives will pertain entirely to this audience. However, if segmentation is used and several audiences are to be targeted, then each will require its own unique set of objectives. Each of these objectives, in turn, will be achieved through the performance of one or more tasks, to be described in the project-planning phase. For each target audience, communicators must ask themselves:

- What behavior change can the intended audience realistically make?
- How willing will the group be to make the change?
- How much time will be required to make these changes?
- Will the achievement of this objective help to reach the emergency preparedness campaign goal?
- To what extent will we be able to measure our progress?

Without objectives, it is impossible to truly measure whether the project has achieved what it had intended to, which makes it extremely difficult to report successes to supporters, partners, and other stakeholders. Often, there is a tendency to see outcomes in terms of what we hope them to be, rather than as they truly are. Setting measurable objectives before the start of the campaign keeps us honest with others and ourselves in determining if the campaign achieved what it set out to do. Therefore, it is vital that these objectives be reasonable and realistic in order to give the campaign a fair chance at being deemed a success (see Sidebar 3.3).

Sidebar 3.3: Setting Reasonable and Realistic Objectives

The NIH recommends that communication campaign objectives be assessed to determine how reasonable and realistic each is concerning the organization's capacity to achieve it. The following is an adaptation from these recommendations:

Be reasonable — Objectives describe the intermediate steps that must be taken to accomplish broader goals; they describe the desired outcome, but not the steps involved in attaining it. Develop reasonable communication objectives by looking at the program's goal and asking, "What can communication feasibly contribute to attaining this goal, given what we know about the type of changes the intended audiences can and will make?"

Communication efforts alone cannot achieve all objectives. Appropriate purposes for communication include:

- Creating a supportive environment for a change (societal or organizational) by influencing attitudes, beliefs, or policies
- Contributing to a broader behavior change initiative by offering messages that motivate, persuade, or enable behavior change within a specific intended audience

Raising awareness or increasing knowledge among individuals or the organizations that reach them is also feasible; however, do not assume that accomplishing such an objective will lead to behavior change. For example, it is unreasonable to expect communication to cause a sustained change of complex behaviors or compensate for a lack of basic emergency services. The ability and willingness of the intended audience to make certain changes also affect the reasonableness of various communication objectives. Your objectives will be reasonable for a particular intended audience only if audience members can make a particular behavior change and are willing to do so.

Be realistic — Once your program has developed reasonable communication objectives, determine which of them are realistic, given your available resources, by answering these questions:

- Which objectives cover the areas that most need to reach the program goal?
- What communication activities will contribute the most to addressing these needs?
- What resources are available? Include:
    - Staff and other human resources
    - Committee members, associates from other programs, volunteers, and others who have the requisite skills and time
    - Overhead resources such as computer time, mailing costs, and printing
    - Services available from another source, such as educational materials available free or at cost and the effort by other organizations willing to help
    - Information about the issue, the intended audience, the community, and media structures, or about available educational materials
    - Budget available to fund the program
    - Time (weeks, months, or years available to complete the program)
- What supportive factors exist (e.g., community activities, other organizations' interests, positive community attitudes)?
- What barriers exist (e.g., obstacles to approval, absence of funding, sensitivity of an issue, intended audience constraints)?
- Which objectives would best use the resources your program has identified and fit within the identified constraints?

Your answers to the last question should become your priority objectives. Sometimes you may feel so constrained by a lack of funds that proceeding appears impossible. An honest assessment may lead you to conclude that a productive communication effort is not possible. However, creative use of the resources already identified may enable you to develop a communication program that can make valuable contributions.

FORM THE PLANNING TEAM AND COALITION

With the setting of goals and objectives, the process of planning the campaign is ready to begin. The project goal and associated objectives are required to drive the generation of numerous individual tasks and actions that will be conducted to achieve the desired outcome. The generation of these tasks, assignation of responsibility, delineation of timelines, and dedication of resources together make up the campaign planning process that is discussed in Chapter 4.

Operational campaign planning begins in earnest with the forming of a planning team or a planning coalition. The myriad tasks that are generated and driven by this planning effort are generally too broad in nature to be effectively conceived of and outlined by just one person, and the value of added perspectives that are gained through the input from a diverse team of planners cannot be underestimated. The utilization of a planning team or coalition also allows communicators to allocate appropriately individual task responsibilities (such as market research or strategic plan development) to those individuals, groups, or organizations most able to handle such tasks — thereby increasing the quality of each distinct campaign function.

The process of forming a planning team begins with the selection or appointment of a project leader, who is normally chosen from within the organization leading the communication effort. The project leader, in turn, leads the development of the full team. Because the quality of the planned campaign will always be reflective of the diversity, knowledge, perspective, and experience of the planning team, the team should be composed of individuals who together satisfy all perceived needs. There are many stakeholders in every communication project, including facilitators, recipients, potential partners, and many others involved in the management of the hazard and its associated risk; each has an important perspective that merits attention and inclusion.

The selected or appointed project leader can begin forming the project planning team by generating a list of all parties and individuals believed to have an interest or stake in the project's outcome. At this point, it must be decided whether the planning effort will remain within the organization or open up to include outside organizations in the form of a coalition. Consideration is given to what each of these organizations and individuals — many of whom are likely to be potential project partners — is able to offer. This includes expertise or knowledge on the preparedness topic, time and effort (volunteer or otherwise), relevant skills, access to contacts and other professional networks, sponsorship, financial support, and much more (see Sidebar 3.5).

If it is determined that a coalition is the wiser choice, then the planning team leader uses the extensive list of potential partners or external planning participants to determine those individuals and organizations whose assistance would be most useful *and* appropriate (keeping in mind the need to maintain representation from the greatest number of stakeholders), and who would be dedicated to the project and work collaboratively with all other team members and partners. Members of this culled list are then invited to join the planning coalition, thereby giving them the opportunity to contribute to the campaign planning effort. Keep in mind that these invitees need not be public education or emergency management professionals. While it is almost always of benefit to include such experts (e.g., firefighters), the coalition may include teachers, respected community officials, business people or leaders, concerned parents and volunteers, and others representative of the target audience. Involving people with different backgrounds and experiences has many advantages, including:

- Access to a wider range of ideas, perspectives, and expertise
- Greater access to the target audience
- Access to additional partners
- A minimized risk of faulty assumptions and methods
- Shared work responsibilities among several people
- Expanded networks of potential contacts, supporters, and sponsors
- Increased access to project funding, labor, and other resources
- Increased message credibility
- Increased message coverage
- Increased levels of trust and attention from target audiences
- Expanded support for priority activities

Coalitions, like partnerships, bring together the knowledge, resources, and commitment of multiple organizations — in this case, members or leaders from different organizations with a stake in emergency preparedness. Ultimately, the attention those organizations pay to the improvement of public preparedness becomes institutionalized for long-term action. For this reason, the strongest potential partners are most likely to be interested in joining a project planning coalition. The NIH recommends using the following guidelines to create a successful coalition (also see Sidebar 3.4):

- Formalize the relationship to create greater commitment — Formal arrangements include written memoranda of understanding, bylaws, mission statements, and regular reminders of the coalition's purpose and progress.
- Make sure that the responsibilities of each organization and its staff are clear — In particular, staff members need to know whether to take direction from the coalition chairperson or from the agency that pays their salary.
- Structure aspects of the coalition's operation — Elect officers. Form standing committees. Have regularly scheduled meetings with written agenda and minutes. Expect and support action, not just discussion, at these meetings. Circulate action items resulting from meetings among coalition members. Establish communication channels and use them frequently.
- Ensure the involvement of representatives who show leadership characteristics, such as the ability to obtain resources, problem-solve, and promote collaboration and equality among members — Members with political knowledge, administrative or communication skills, or access to the media and decision-makers are also valuable.
- Create and reinforce positive expectations by providing information on the coalition's progress — Optimism and success sustain member interest.
- Formalize accountability and develop criteria for judging whether coalition members are honoring their commitments.
- Be flexible — Losing prospective partners can limit a program's effectiveness.
- Provide training to help members complete their tasks — For example, coalition members may need training in how to be effective advocates for your program's issues.

- Give members a stake in the coalition and an active role in decision-making.
- Seek external resources to augment member resources.
- Evaluate the effectiveness of the coalition periodically and make necessary changes — This should include process evaluation of the coalition's functioning and assessment of the coalition's impact on the health problem being addressed.

Tucker and McNerney describe four kinds of coalitions. The FEMA Emergency Management Institute provides examples of how each may relate to a public disaster preparedness education campaign:

1. Representatives of different groups who have grown weary of costly confrontation. They need to build consensus, using a specific issue as common ground. Example: *Various voluntary community groups, competing to garner the most support from private and corporate donations, might instead combine efforts on the topic of community preparedness, and share otherwise limited resources to achieve a wider level of preparedness.*
2. Representatives of different groups who, although of different missions or opinions, realize that they share a common perspective on a specific issue. Example: *Insurance companies and emergency preparedness authorities might have different goals (selling insurance to cover disaster claims versus providing adequate responses in a major population emergency), but both will benefit if the community is prepared (less damage/less claims, and less dependence on limited police/fire/ambulance resources in a disaster).*
3. Representatives of groups with varied goals and perspectives who are more likely to be sensitive to the specific point of view. Example: *A municipal health department with oversight for ambulance, public health, hospital, and clinic services might be sensitive to community emergency preparedness because the public's health needs in an emergency will affect the agency by demanding a response for many who, if not prepared, might otherwise become injured, need medical services, or become homeless.*
4. Representatives of varied groups might share a position that already has widespread acceptance. Example: *Businesses, service organizations, and governmental agencies all might climb on the bandwagon to publicize a smoke detector installation campaign in residences. Such efforts can sell detectors, reduce overall insurance liability, limit*

loss of life and property, and make the jobs of fire departments safer and more effective.

Sidebar 3.4: Six Steps to Forming a Coalition

FEMA describes six steps that may be used to build and organize coalitions. The following is an adaptation of these steps:

1. **Develop a position:** Evaluate the importance of the preparedness issue, who is driving it, and who is likely to be affected or think they will be affected by it. Create a position that will benefit your organization's success with the program, as well as those you are trying to educate.
2. **Create a strategy for pursuing your position:** Questions to consider include:

   How can you accelerate the opportunity to enhance community preparedness?
   How can you make adjustments to adapt to the community's need or capitalize on a given trend?
   Is there a mutual benefit for the community and for businesses to promote preparedness education and activities?
   Can you get a major organization to back the concept or offer incentives for those taking specific training or actions?
   Is there a sense of ownership among participants?

   A key strategy is to help develop a sense of ownership of specific outcomes among coalition participants. The partners must understand and embrace these outcomes as goals they want to achieve.
3. **Identify coalition participants:** For each organization considered, planners should think about the following questions:

   What are their positions?
   How credible is the organization/individual with other organizations or individuals?
   Will they want to be official sponsors of the effort?
   Who are the leaders of those organizations?
   Will they participate themselves or will they recommend someone who will?

How can I work with others within those organizations or who have influence on them to help shape their opinion and see the preparedness point of view more favorably?

4. **Conduct research:** Try to get a baseline on the level of community preparedness before beginning a campaign. This evidence is used to support the need for your preparedness education programs.
5. **Organize your meeting:** Try to find common ground among invited coalition members. Where do we agree and disagree? Where can we work together? Try to identify common philosophical values and look for misperceptions and unrealistic expectations. With groups of differing perspectives, building consensus will not be easy. Participants will need to work through the process, so everyone must be able to let his or her positions be known.
6. **Plan the campaign.**

Planners should be aware that there can be drawbacks to including partners in the planning process and in the facilitation of the campaign itself, but awareness of these issues can help them to be minimized. Drawbacks to including partners can include:

- Identifying partners, persuading them to join your efforts, waiting for them to make a decision, training them in the relevant issues, and coordinating with the additional team members often serve to increase the burden of time involved in planning.
- The different wants, needs, perspectives, experiences, capabilities, and ideals of each partner can require that the nature of the campaign be altered to ensure that all partners are satisfied.
- Partner organizations may try to use the program for their own needs, or take credit for the program's successes beyond the contribution they provided.
- Staffing problems, funding shortages, or mismanagement in partner organizations can all lead to delays, mismanagement, or complete failure of the campaign.

For these reasons, planners must be sure that they are prepared to work with whatever partners they recruit, and that those partners are willing and able to do so (and for the right reasons). They must have a solid understanding about how flexible they are willing to be with the campaign to meet the wants of their partners, and how much support they are willing and able to provide them if and when they require it.

Whatever the makeup of the team that results from this effort, it must consist of enthusiastic supporters who can help plan and promote the public education effort. Ideally, representatives from the target audience are included as full members or in advisory roles, to verify assumptions and provide subject matter expertise. It is equally important to involve community organizations that typically work with the target audience, primarily to prevent redundancy of effort, provide the mutual benefit of collaboration, and allow for the sharing of ideas and experiences (as well as access reputation and trust they enjoy among message recipients). While it is important that at least two or three people in the core planning team are able to provide leadership and continuity throughout the planning and implementation effort, it is okay if some planning team members rotate in and out as appropriate (with some available only for the initial planning meetings and others helping out only during implementation).

The target audience profile will heavily influence the team's membership. For instance, businesses and industries can serve as effective conduits for sharing disaster information among their employees, if their employees are among those targeted. Businesses typically have phone trees established and methods for disseminating information to staff. Furthermore, many companies have already established disaster education programs that the planning team can tap into (such as emergency drills and annual training sessions). Government agencies and other community organizations are also important partners in public education. School systems, for example, can help educate children and young adults about hazards in their areas and appropriate preparations and response measures. Many government emergency management offices have already assessed the hazards in the community, and may have even developed guidance on how to educate their constituents.

One of the greatest benefits of including a wide range of stakeholders is that each will provide important input into the process specific to his or her individual perspective, and will likewise become a vocal advocate for preparedness in general and the project at hand. The following are examples of stakeholders that may be included in a public education planning effort:

- Emergency responders (fire, police, EMS) and emergency managers
- Local, state, and federal governmental agency officials

- Private sector/business/industry leaders
- Volunteer organization representatives
- Community and faith-based organization leaders
- Elected officials
- News media representatives (television, newspapers, radio, Internet)
- Representatives from the target audience
- Educators and school administrators (from schools, colleges, and universities)
- Concerned individuals
- Civic and business organizations
- Businesses
- The Chamber of Commerce and the area Council of Governments
- Local community centers
- Religious organizations
- Youth clubs
- Women's clubs and organizations
- Trade enterprises and associations
- Banks and credit unions
- Health centers, hospitals, or clinics
- Sport clubs
- Libraries, cinemas, theaters, or circuses
- Utility companies
- Red Cross chapters

Encouraging Partners to Join the Planning Team

The inclusion of partners can be the most effective way to expand the scope and reach of the campaign. In exchange for the benefit of participating, partners bring to the project skills, labor, equipment, audience access, credibility, materials and supplies, space, experience, and much more (see Sidebar 3.5 for examples of benefits gained through partnership). How and when partners are identified and approached is always at the discretion of the organization conducting the campaign. In some cases, if your organization has little or no experience with public education, it may be preferable to include partners from the very beginning of the planning process before the problem is fully defined or the audience is profiled. However, in most cases it is preferable that there be structure to the project so that partners may know in what they are agreeing to participate.

Sidebar 3.5: Partner Organization Contributions

Partnering organizations bring to the program a much wider range of skills, abilities, and resources than are possessed by any single organization. The addition of these attributes allows planners to consider many more options than they otherwise could. The skills and resources a partner organization can bring to a public disaster preparedness education campaign might include:

- Graphic design software, skills, and equipment
- Printing materials and equipment
- Advertising space or time
- Endorsement
- Sponsorship or inclusion at events
- Specialized knowledge of or access to the target audience
- Specialized skills
- Additional people to communicate the public education message
- Space to hold events and equipment required to do so
- Food, drinks, and other supplies to draw people to events
- Storage, transport, or distribution of materials
- Experience with the hazards addressed
- Increase in the number of messages the program is able to transmit
- Training resources
- Expanded support for your organization's priority activities

With luck, partners may jump at the chance to participate in your project as proposed. This is most typical for organizations that regularly work with the target audience and when these partners are required to dedicate little or no resources of their own. However, most potential partners will need convincing before they agree to join. The planning team leader must therefore be able to present to these organizations and individuals the benefits each stands to gain through their participation. Such benefits could include:

- The opportunity to share credit for success
- Membership and participation in a forum whereby community problems are discussed, addressed, and resolved

- The opportunity to foster good community relations
- Increased awareness of the hazards faced by community residents and businesses
- The opportunity to improve the working relationships between government and civil society
- Local and collective ownership for the resolution of community problems
- Increased visibility and credibility in the community
- The opportunity to build organizational capacities and other skills
- Networking opportunities
- Increased positive media coverage, perceived credibility, and community visibility
- Access to data and experience
- Assurance of message accuracy

Among the wide variety of options for partnership opportunities, there are subgroups within this body that can be approached for specific reasons. Major subgroups of partners in the community include the following:

- **Local businesses** — There are two primary reasons why local businesses participate in community projects like a public disaster preparedness education campaign. The first is that these businesses depend upon their good reputation among community members. A project such as this can significantly raise their profile within the community, as its goal is to reach as many members of an audience as possible. The second reason is one of corporate responsibility. Many businesses feel they should give back to the communities that make their success possible. Public education projects in general can require a significant amount of skills, equipment, and materials, all of which are provided by different members of the local business community. By partnering with these organizations, it is possible to acquire access to these skills, equipment, and materials as an in-kind donation in exchange for the positive publicity that comes through shared project credit. Through creative planning, it is often possible to gain these items through no extra cost on the part of the partners. For instance, grocers, hardware store owners, and other merchants in your community may be invited to put preparedness messages on shopping bags, store windows, or marquees outside their stores (see Figure 3.3). Local businesses may agree to work with you to set up displays inside their stores featuring key items shoppers

FIGURE 3.3 Tulsa, Oklahoma, Mayor's Citizen Corps received donated space on food tray liners from the McDonald's Corporation for printing emergency preparedness education materials for disaster education. (Image used with permission by Tulsa Mayor's Citizen Corps and the Oklahoma Department of Emergency Management.)

might need as they put together a family emergency supply kit. Local merchants might also be interested in sponsoring contests on preparedness in the schools and donating prizes for kids. Other examples of opportunities for partnership roles for local businesses include:

- Local utilities (telephone, water, electric, gas) can include emergency preparedness and mitigation messages or literature in customer bills or newsletters.
- Businesses can be encouraged to distribute disaster preparedness information to employees, and hold preparedness workshops with employees and their families during business hours.
- Graphic design, printing, and other businesses with related capacities can design and print hazard and basic preparedness fact sheets and brochures.
- Local organizations and businesses can donate space upon which a poster may be hung or where brochures may be distributed to customers.
- Businesses with a large customer base may allow access to their customers by permitting the organization to host in-store workshops or other events.
- Restaurants or fast-food chains may want to donate food or refreshments at events.

- **Community organizations** — Community organizations include youth clubs, Red Cross chapters, NGOs, law enforcement organizations, women's groups, veteran's groups, religious organizations, and others. While these organizations are rarely able to provide financial assistance to the project, they can often provide two equally important resources: people and credibility. Community organizations tend to have deep roots in the community, and may enjoy an even higher level of recognition and trust among community members in general or the target population in particular. Members of these groups work to help their community in a variety of ways, and when the public education project is compatible with those goals, they are likely to pay close attention. By accessing the networks each of these groups has established, your team can greatly extend the reach of its disaster awareness and preparedness message. Some ideas of partnership roles for community organizations include:

   - Organizations that hold meetings with members of the community or some other target population can present disaster preparedness and mitigation workshops at one of these meetings.
   - Organizations that hold periodic fairs, festivals, or other events can include a segment or booth on disaster preparedness.
   - Organizations that work with the target population can incorporate disaster preparedness into the services that they regularly offer.
   - Organizations with a wide volunteer base can offer the time or skills of their volunteers for the planning or facilitation of the campaign.
   - Organizations may contribute space to hold meetings, store equipment and supplies, or host people working on the project.
   - Organizations may provide equipment such as computers, software, printers and plotters, audio and video recording and display devices, and so forth.
- **Public safety organizations** — The various public safety organizations that operate in the community already have a vested stake in the preparedness of the community populace. These organizations may even have public education campaigns in place and operational. By working with them, planners will tap into a great amount of community-specific knowledge regarding hazards, solutions, and audiences. Additionally, the credibility these organizations hold within the community is rarely surpassed. The range of contributions these organizations may provide is limited primarily by the available time of their members, and their resources. However, as the vast majority of public safety organizations depend upon financial contributions from the community, they are likely to participate in any event or project that can increase their likelihood of receiving grants or donations. Creativity on the part of the planning team, and the ability of the planning team members to tap into their very limited time, will be the key to their involvement. Some ideas for partnership roles for public safety organizations include:
   - Firefighters, policemen, emergency managers, or EMTs may present public safety campaign messages through presentations at schools, civic group meetings, or other public venues (see Figure 3.4).

FIGURE 3.4 Emergency services departments can combine annual fundraising efforts with public education campaigns by holding family preparedness fairs that encourage children and families to prepare by showcasing the resources and capabilities of the department (Coppola, 2008).

- Public safety organizations can offer sponsorship to a campaign, thereby lending a great deal of credibility to the message.
- Public safety organizations may have valuable information that can be used to define the hazard, possible solutions to the problem, or information specific to the target audience.
- Public safety organizations may have unique access to members of the target audience, or authority to conduct specific tasks (such as home safety inspections).

- **Partnering with schools**—Schools provide quite possibly the best access to any target audience that includes children or families.

FIGURE 3.5 A fourth grader at Painted Rock Elementary School dons firefighter gear during a FEMA for Kids presentation in Poway, California, February 13, 2008 (John Ashton for FEMA, 2008).

The environment found within schools is perfectly suited to foster learning, so any message conducted within this setting is likely to benefit (see Figure 3.5). In addition, schools already participate in many other public health education campaigns, so the structures are in place to transmit the disaster preparedness message. Through school-based associations, including parent–teacher groups or sports teams, planners can extend their message reach. Students are often enthusiastic about participating in public education projects and can be relied upon to echo preparedness messages throughout the community. Local merchants might even be more likely to participate if they know that schools are involved, because of the wider customer base that is reached and the credibility that is gained. Examples of partnership roles for schools include:

- Schools can hold poster competitions, with prizes (donated by local businesses, for instance) for contest winners.
- Schools can incorporate a disaster preparedness workshop into their regular parent and teacher meetings.
- Schoolteachers can incorporate a disaster preparedness message into their lesson plans.

  - Schools can assign at-home projects that involve family disaster preparedness activities.
- **Target audience leaders** — Oftentimes, planners are not necessarily experts on the issues faced by the special populations their public education efforts are targeting. By involving leaders and members of these target groups in the planning process, they are not only ensuring that all of the important topics and issues are at least considered, they are also helping to establish trust and buy-in among the targeted individuals. Examples of partnership roles for target audience members and leaders include:
  - Audience members and leaders can assist with the development of a target audience profile.
  - Audience members and leaders can participate in the pretesting of messages and materials before they are distributed among the greater target audience.
  - Target audience leaders can endorse the project, lending significant credibility to the campaign.
  - Target audience members and leaders can help the organization to assess the progress of the campaign and to make required adjustments.

The Media as a Partner

In most public disaster preparedness endeavors, a partnership with the news (mass) media will be paramount, offering quite possibly the most important and influential contribution an organization can gain through its partnership endeavors. News media outlets, which include television, print (newspapers), radio, and Internet-based companies, already play a significant role in disaster and emergency management both before and after disasters occur. The media, for instance, are lauded for the valuable service they perform during the initial critical moments of a disaster when emergency response efforts are first mobilized, and media organizations serve to transmit warnings, evacuation orders and instructions, the location and availability of medical care and shelter, and where to go for more specific information.

In the preparedness phase of emergency management, the primary public education tasks assumed by the media are very similar (if not identical) to what you are likely to be performing yourself, including raising citizen awareness to the presence of hazards and providing information to those citizens regarding prevention or protection measures. The media have established themselves within modern society such that, as a general population, citizens turn to them more than any other source to obtain information, including that which relates to hazards (Walsh, 1996). FEMA mitigation specialists even go so far as to claim that the media role in community and citizen preparedness is critical if such efforts are to succeed (FEMA, 1998).

The media role should not be overestimated, however, as it is not absolute. While it has been found that personal preparedness is most likely to be undertaken by people attentive to the news media, this tendency is usually accompanied by other behavioral characteristics that support preparedness actions and attitudes (e.g., personal experience and expendable income). In this sense, the media role should be seen as an important *supplemental* component in your preparedness campaign, not the answer to its problems. Campaigns that rely too heavily on the media without addressing the reasons why people do not or are unable to prepare will not succeed. Other specific problems associated with media participation in public disaster preparedness education campaigns include the following:

- Although the media is effective at raising awareness about issues and communicating degrees of urgency, they often avoid contributing solutions to problems.
- The media is often unable to educate the public about risk in such a way as to give citizens an accurate perception of personal vulnerability.
- The media often speaks to a general audience rather than addressing the specific needs of a more focused target population.

The mass media is diverse in character, utilizing print, broadcast, and other methods of transmission, and existing at the local, regional, national, and international levels. It is a well-established institution operating in a predictable manner. With the right strategy, planners can tap into these resources and channels to reach wider audiences easier and quicker than they ever could by other means. FEMA proclaims in the publication *Project Impact: Building a Disaster Resistant Community,*

> You will want to target print, radio, and television outlets at planned intervals with your messages. As gatekeepers to your community, the media affect and shape our opinions and our behavior. They influence our preferences and our choices. By encouraging reporters to write or broadcast your messages, you will generate awareness and interest.

FEMA claims that a targeted, comprehensive media list, containing all of the important and relevant media outlets that reach the target audience, is the "most essential tool of any successful media campaign." The planning team can create a media list through cooperation with local government agencies and other organizations that maintain regular contact with members of the mass media. Sources of greatest value will likely be those that cover community affairs, natural disasters, or the metro desk, for example. Outlets that should be included in the media list are found in Sidebar 3.6.

You must always remember that members of the media are not public education experts, and their goal is not to inform the public but rather to increase the ratings of their media outlet. Therefore, careful attention must be paid when fostering partnerships with the media to ensure that their inclusion does not backfire and end up hurting rather than helping your cause. Peter Sandman, an acclaimed risk communication expert, describes 11 ways for those performing preparedness education to help reporters understand the technical aspects of a story or message. They are (Sandman, 1992):

1. Don't assume knowledge.
2. Guide the interview.
3. Avoid jargon.
4. Simplify content.
5. Anticipate problem areas.
6. Provide written back-up information.
7. Be alert for signs of confusion.
8. Check for understanding.
9. Suggest other sources.
10. Offer to look at a draft or check quotes.
11. Encourage specialized reporting.

Media partners can disseminate preparedness messages through various means, which may include articles, feature stories, editorial coverage, or donated advertising space. In its *How-To Guide for State and Local Mitigation Planning*, FEMA describes various ways in which a planning team can work directly with the media to promote the risk communication messages it has developed, such as:

- Include a special insert in a local newspaper.
- Broadcast public meetings on a local access channel or through public service announcements.

- Produce a video to be broadcast on local access channels.
- Use news releases or information contained in press kits to create feature stories or reports. Press kits are folders summarizing the key information about your goals and actions, information that helps to pique interest in your program, and information that provides reporters with accurate details about the hazards and what can be done about them.
- Announce an upcoming meeting or event.
- Attend a meeting or event to highlight your cause.
- Provide viewers with contact information or other important data that will help them to locate preparedness instruction and information.

The planning team can also contact local broadcasters and offer interviews with disaster safety experts. Television stations frequently need guests to fill slots in early morning or weekend shows. Radio talk show hosts may welcome the chance to interview an emergency preparedness expert provided by the team, because the topic is always timely. The team can also arrange visits to the editorial department of local newspapers to gain print coverage. Experts could include representatives of the local or regional Red Cross, Salvation Army, emergency management office, National Weather Service, or fire department rescue team. The team may also want to have the expert write an opinion piece or a letter to the editor to be submitted to the newspaper.

FEMA provides some caution for working with the media. It states,

> [w]hile the media is a good source for getting information to the public, you do have to be careful. Sometimes the media can distort the information you give them or give it a different spin. The media likes attention-grabbing headlines so they may try to make your plan controversial in some way. You should work on establishing an honest, working relationship with a local reporter so that each of you has someone to turn to when you need to gather or provide information to the community.

Sidebar 3.6: Media Checklist

Media lists should include:

- Newspapers (dailies, weeklies, monthlies, college/university papers, and community newsletters)
- City and regional magazines

- Local trade and business publications
- State bureaus of national wire services, such as the Associated Press (AP), Reuters, and United Press International (UPI)
- Local radio and television stations (including college/university networks)
- Local cable stations
- Public broadcasting stations (which may have community affairs programming)
- Public information officers at military bases, if applicable (many military housing areas have broadcast stations and newsletters that may reach the entire families of service members)

Regardless of the medium, for the most part your media list will consist of the following types of reporters:

- Metro desk/city reporters — interested in news around town
- Public affairs reporters — interested in civic and legislative issues
- Business reporters — interested in hard news involving regional business, local economy, and economic/community growth (e.g., impact on sales, environment, address changes)
- News assignment editors
- Public service announcement directors

Source: Federal Emergency Management Agency (FEMA). 1998. *Making Your Community Disaster Resistant: Project Impact Media Partnership Guide*. Washington, D.C.: FEMA.

DRAWING UP PARTNERSHIP PLANS

Projects run more smoothly when everyone involved assumes his or her role with an accurate impression and full understanding of what he or she is expected to contribute. This certainly holds true with partners. Partnership plans and agreements can be drawn up to manage expectations. These agreements not only ensure that partners understand what role they are expected to play, they also help to prevent them from overstepping their bounds and taking too much control of the project. Project managers can easily lose control and ownership if any one organization begins working far beyond the expectations of the original partnership agreement, and the partnership plan is a good reference that keeps these risks in check.

PROJECT MANAGEMENT

All complex projects require project management. The project management effort depends upon a project manager who is able to direct all players and resources according to the tasks required and the timetable desired. Project management must cover the project from planning and development, through implementation, to evaluation. The most effective method of maintaining project management is to create a visualization of each of these elements. Identifying and describing all tasks at the start of the project is very effective in ensuring that the project does not run into hidden or unexpected overruns in time or cost.

The tasks involved in a public disaster preparedness education campaign are not conducted in a purely linear fashion. Often, different members of the campaign conduct many tasks at once. To increase the efficiency of time, labor, and resources, and to minimize the time required for project development and campaign facilitation, the project manager will need to be aware of each task and its current status. This is usually conducted with a task list.

Task lists are most effective when they are ordered by start date, with the anticipated time to conduct the task noted. An effective means to illustrate the task list is a cascade Gantt chart, so called because of its waterfall-like appearance. An example of a Gantt chart is provided in Figure 3.6. A timeline is drawn to each task (respective to all other tasks), responsible individuals or organizations are assigned, and resources are noted.

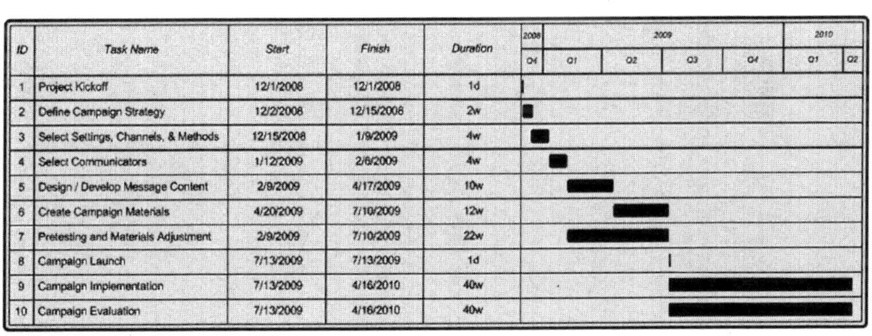

FIGURE 3.6 Example of a Gantt chart used to organize and illustrate the tasks of a preparedness campaign.

The project manager, who is responsible for assigning people and organizations to tasks, can use a Gantt chart or similar illustrative task list to ensure that no one is double booked, and that all resources are identified, acquired, and assigned to no more than one task. These figures also make it easier for the project manager to ensure that tasks are being conducted on time and in the order in which they are needed. While by no means required, there exists off-the-shelf software, like Microsoft Project or EasyProjects.net, which guides the project management process. Whatever method is used, it should be flexible enough to allow for changes as the project progresses, and the true completion dates and costs become apparent.

CONCLUSION

Emergency preparedness public education campaigns may be complex, but with proper planning, staffing, and funding, and with a comprehensive understanding of the target audience, they can be very successful. Campaign planners are wise to dedicate as much time as possible to the planning phase — often as much or even more time as the actual campaign itself. The next two chapters describe the process by which these plans become practice, and the public is educated about the actions it may take to reduce its hazard risk.

REFERENCES

American Red Cross. 1992. *Community Disaster Education Guide.* Washington, D.C.: The American National Red Cross.

Bernstein, D. A., A. Clark-Stewart, E. J. Roy, T. K. Srull, and C. D. Wickens. 1994. *Psychology*, 3rd ed. Boston: Houghton Mifflin.

Buss, D. M., K. H. Craik, and K. M. Dake. 1986. Contemporary worldviews and perception of the technological system. In *Risk Evaluation and Management.* V. T. Covello, J. Menkes, and J. L. Mumpower, Eds. New York: Plenum.

Coppola, Damon. 2006. *Introduction to International Disaster Management.* Burlington, MA: Butterworth Heinemann.

Dake, K. (1991). Orienting dispositions in the perceptions of risk: An analysis of contemporary worldviews and cultural biases. *Journal of Cross-Cultural Psychology*, 22, 61–82.

Dake, K. 1992. Myths of nature: Culture and social construction of risk. *Journal of Social Issues, 48,* 21-37.

Federal Emergency Management Agency (FEMA). 1998. Making Your Community Disaster Resistant: Project Impact Media Partnership Guide. Washington, D.C.: Author.

FEMA. 1998. Project Impact: Building a Disaster Resistant Community. FEMA Publications. Washington, DC.

FEMA. 2002. How-To Guide for State and Local Mitigation Planning. FEMA Publications. Washington, DC.

FEMA. 2004. Hazards Risk Management. Emergency Management Institute. Higher Education Program. http://training.fema.gov/EMIWeb/edu/hram.asp

Jasper, J. M. 1990. *Nuclear Politics: Energy and the Slate in the United States, Sweden, and France.* Princeton, N.J.: Princeton University Press.

Jenkins-Smith, H. C. 1993. *Nuclear Imagery and Regional Stigma: Testing Hypotheses of Image Acquisition and Valuation Regarding Nevada.* Technical report Institute for Public Policy, University of New Mexico: Albuquerque.

National Cancer Institute. 2004. *Making Public Health Communications Work: A Planners Guide.* Washington, D.C.: National Institutes of Health..

Ropeik, David. 2002. "Fear factors" in an age of terrorism. October 15. http://www.msnbc.msn.com/id/3077306/

Sandman, Peter M. 1992. Helping Reporters Understand a Technical Story. Conference Handouts. http://www.psandman.com.

Slovic, Paul. 1987. Perception of risk. *Science,* 236, 280–285.

Slovic, P. 1999. Trust, emotion, sex, politics, and science: Surveying the risk-assessment battlefield. *Risk Analysis* 19(4), 689–701.

Slovic, P., B. Fischoff, and S. Lichtenstein. 1979. Rating the risks. *Environment,* 21, 14–20, 36–39.

Slovic, P., B. Fischhoff, and S. Lichtenstein. 1980. Facts and fears: Understanding perceived risk. In *Societal Risk Assessment: How Safe Is Safe Enough?* New York: Plenum.

Tucker K. and McNerney S. L. 1992. Building coalitions to initiate change. Public Relations Journal 48(1) 38-40.

Walsh, James. 1996. *True Odds: How Risk Affects Your Everyday Life.* Santa Monica, CA: Merritt Publishing.

4

Step 2
Develop a Campaign Strategy

INTRODUCTION

The second step in the public preparedness campaign process focuses on the development of a campaign strategy and a communication plan. In this step, the planning team works together to determine how they intend to communicate the risk reduction solutions identified in Step 1. Campaign strategy development, which draws primarily from the market research conducted in preparation for the communication to be conducted, begins with the selection of appropriate communication settings, channels, and methods, each of which is described in this chapter. The communicators who will be transmitting the messages must be identified and recruited, and the timing of communication events and message transmissions must be planned. Perhaps the most complex element of the communication effort — the development of campaign messages and materials — follows. Finally, a communication plan is created, and staff is trained, thereby paving the way for the actual communication efforts.

PROJECT KICKOFF

Projects that rely upon a cooperative effort almost always stand to benefit from a milestone event marking their commencement, and public emergency preparedness efforts are certainly no exception. It is common practice for a project manager to hold a project kickoff meeting that serves to (among other things, of course) give all project stakeholders a sense that planning and operational efforts are officially under way. In order to generate excitement among participants, this event is generally celebratory in nature, is oftentimes catered with food and drinks and adorned with banners and other decorations, and may even include prizes and incentives like logo-bearing, project-related giveaways that help to raise excitement and morale among all participants. With proper foresight and planning, these important events can also provide much more than simple ceremony, including public recognition and the development of strong partner relationships from which the public education effort will grow. Moreover, for most projects the kickoff meeting is the first opportunity for all stakeholders to gather, giving the project manager a controlled opportunity to bestow a first impression of trust and confidence to all participants.

The adage "the more the merrier" holds true with project kickoff meetings. In fact, attendance should be inclusive of everyone involved in the planning and operation of the project, or at minimum a representative from each group and organization if the project effort is an especially large one. This could include:

- The project manager
- Representatives from the lead organization
- The full project planning team
- Members of partner organizations who might not be involved in the planning effort
- Businesses that are lending support to the campaign
- Project sponsors
- Organizations or individuals who will provide training to team members
- IT and other support staff
- Target audience leaders
- Consultants and contractors

The kickoff meeting serves as a venue for several important project tasks that must occur early in the project. However, for any of the individuals or groups that are participating in the project for the first time, the amount of uncertainty will be high, as will be the learning curve of knowledge and understanding. The project manager can help to ensure that the meeting is effective in addressing these concerns by using a structured organization, guided by an agenda provided to all participants in advance. Kickoff meeting participants who are part of the project planning team or coalition should be provided (if they have not already) with a packet containing information about the project and all participants and partners (including names, phone and fax numbers, and email and mailing addresses). Tasks that should be covered in the meeting include (but are not limited to):

- Introduction of all participants to one another
- Sharing and verification of all contact and communication information
- Setting in motion the official start of the project
- Communication of project purpose, goals, scope, and timelines
- Presentation on the intended audience, their vulnerabilities, and proposed solutions
- Communication to all partners of their roles, responsibilities (tasks), and resources
- Voicing of all partner assumptions and concerns
- Discussion of the project budget
- Discussion of project risks
- Discussion of training needs

The project manager can commence this meeting with a prepared motivational statement, and provide all facilitation of the meeting as dictated in the agenda. If there are any conflicts between partners, concerns of participants, or other input or suggestions, this is a great opportunity to identify and manage them early on, before actual project development has begun. Again, it is the impression of organization and control that are of the utmost importance at this initial meeting.

THE CAMPAIGN STRATEGY

In order to begin developing the campaign in a structured, organized fashion, a campaign strategy must be drawn up. A campaign strategy profiles the processes and procedures by which a communication program intends to achieve its risk reduction goals. The strategy, which is created in consideration of resources available to communicators, restrictions in time, and logistical capacity, outlines the guiding principles behind all of the campaign's messages, materials, events, and activities. Ideally, the strategy provides readers with sufficient detail about the target audience's profile, the hazard risk being addressed, the desired behavioral changes and benefits expected as a result, and the actions that can be performed to bring about those changes, to ensure that all program elements are compatible with the campaign goals and objectives (and with each other). It gives the program's creative staff the direction they need to develop the messages and materials that will convey the campaign messages. It must be developed in the best interests of the target audience, and should be favorably viewed by that audience should members be given a chance to examine it. The NIH details the components of a campaign strategy as follows:

- A definition and description of the intended audience.
- A description of the action the intended audience members should take as a result of exposure to the communication.
- A list of any obstacles to taking action (including external influential variables — see Sidebar 4.1 and Sidebar 4.2).
- The perceived benefit (among target audience members) of taking the action.
- A description of the support that will make the benefit, and its ability to attain it, credible to the intended audience. (In this case, support is something that can be provided through hard data, peer testimonials, demonstrations, or statements from organizations the intended audience finds credible.)
- The settings, channels, and activities that will reach intended audience members, particularly when they will be receptive to or able to act upon the message.
- The image your program plans to convey through the tone, look, and feel of messages and materials. NIH writes that the goal should be to convey an image that: (1) convinces intended audience members that the communication is for them; and (2) is culturally appropriate. For instance, printed materials convey image through typeface, layout, visuals, color, language, and paper stock used. Web materials convey image through design, typeface, color, layout, and ease of use. Audio materials convey image through voices, language, and music; in addition to these details, video materials convey image through visuals, characteristics of the actors (including their clothing and accessories), camera angles, and editing.

The following is an example of a campaign strategy:

Strategy: Hold flood mitigation training sessions for property owners in the National Flood Insurance Program (NFIP) Special Flood Hazard Area (SFHA).

Intended Audience: Low- to middle-income business owners and homeowners whose properties lie at or below the floodplain mitigation goal of base flood elevation (BFE) plus 2 feet.

Objectives: (1) to understand the mitigation options available that would make their structures resilient to flood risk; (2) to understand the NFIP and the options for flood insurance; (3) to obtain and understand a FEMA elevation certificate; and (4) to select and invest in flood mitigation options that provide the greatest increase in protection for their structures for the money spent.

Obstacles:

- Unavailability of financial resources
- The lack of technical knowledge required to fully understand certain mitigation options
- Misperceptions of risk (feelings that "it can't happen to me")
- Undesirable aesthetic changes to homes or businesses
- Loss of use of certain parts of the structure
- Low self-efficacy
- Lack of time
- Competition with other structural improvement needs

Key Promise:

- If I maintain adequate flood insurance for my flood-prone structure, my losses will be covered in the event of a disaster, and I will not be disqualified for future emergency assistance.
- If I mitigate my structure to at least BFE + 2 feet, my flood risk will decrease significantly and I will enjoy much lower flood insurance premiums.
- By mitigating my structure from flood risk, I will be much less likely to suffer the long-term negative effects of a flood disaster, will be displaced from my home for a much shorter period of time, will suffer much less of a business interruption, and will have far fewer financial and physical losses.

Support Statements:

- Every year, flooding causes more than 90% of the disaster-related property damage in the United States, and accounts for more than 75% of all presidential disaster declarations.

Floods are one of nature's most powerful destructive forces, and they cause more damage than all other forms of disaster in the United States combined.

- If your home or business has been threatened or damaged by a flood or severe storm, and you have not made any changes to the structure that protect it from water damage, you are probably still at risk for similar or worse consequences when flood events occur in the future.
- Through engineering, architecture, and sheer determination, it is possible to reduce, or even eliminate, your chances of falling victim to flooding.
- The NFIP is a federal program that enables property owners to insure their property against flood losses, most of which are not covered under regular business, renter, and homeowner insurance policies.
- Structures that are mitigated from flood risk have a much higher resale value than those that are at risk of flood damage.
- Structures located within the NFIP special flood hazard area are legally required to have flood insurance if they are purchased with a federally backed mortgage.

Tone: Urgency without fear; empowerment.

Channels:

- Interactive training sessions of no more than 10 attendees per session, with PowerPoint presentation, Q&A opportunities, and speakers from the NFIP, the County Floodplain Administrator's office, and community organizations
- Descriptive, easy to understand flood mitigation guidebook
- Flood mitigation fact sheets
- Flood mitigation helpline

Media:

- Local television public service announcements (PSAs)
- Radio PSAs/live announcer scripts
- Posters
- Newspaper ads
- Billboard and metro transit ads
- Website
- Articles in local newsletters

Settings:

- Rush hour commute
- Evening news

- Sunday real-estate section of the newspaper
- Day, evening, and weekend training sessions

Creative Considerations:

- Solutions must be adaptable to local needs and capacity.
- Certain mitigation measures may be regulated by floodplain management ordinances.
- Some property owners may have taken mitigation measures that brought their property into compliance with older flood maps, but under new BFE measures, they are no longer compliant.

This example shows a brief strategy statement. Depending upon the complexity of the campaign, strategy statements of greater detail and length may be warranted to account for multiple projects being performed under the same program umbrella. They may also contain additional background information about the hazard, its risk, the affected target population, and other factors that will assist planning team members in developing the campaign along a unified front.

Sidebar 4.1: Risk Communication Obstacles and External Influential Variables

Risk communication is as recognizable for its importance as it is for the complex challenges associated with it. While the World Bank claims that "awareness programs addressing existing hazards and physical and social vulnerabilities are often central to social risk reduction" (World Bank 2002), the National Research Council (NRC) Committee on Risk Perception and Communication writes, "risk messages are difficult to formulate in ways that are accurate, clear, and not misleading" (NRC 1989).

Countless risk communicators have experienced project failure caused by obstacles they could not overcome, for which they did not plan, or of which they were simply unaware. These obstacles have included internal and external political affairs, socio-cultural issues, and economic constraints of both communicators and recipients. In addition, while some obstacles may be obvious and even well documented prior to initiation of the campaign, many may not appear until well after the campaign has begun implementation. Communicators must always understand that, even with the best-laid plans, the effectiveness of risk communication can and often does fall short of the communicator's expectations because of the presence of obstacles (Morgan, Fischoff, Bostrom, and Atman, 2002). If obstacles are not considered in the planning stages, the communicators may face insurmountable problems in the execution of their project. It is therefore vital that risk communicators prepare themselves to identify and manage obstacles as they arise.

Successful risk communication tends to be highly situation-dependant, and practitioners who succeeded in reaching target audiences with their intended messages and effecting change are those who have sufficiently identified and mitigated for situational obstacles. Obstacles differ from audience to audience, and community to community. As such, educators working in rural communities may encounter obstacles not typically encountered in urban communities, for example. Some of the obstacles that should be factored in to risk communication projects are reviewed in the following.

- **Literacy and Education** — According to a study conducted by the United Nations Educational, Scientific, and Cultural Organization (UNESCO), there are an estimated 862 million illiterate adults throughout the world, approximately two-thirds of whom are women (UNESCO Press, 2002). Illiteracy severely limits how risk communication can be conducted. There are countless ways a message can be distributed through written media, such as leaflets, newspapers, billboards, and informational booklets, to name a few. However, these tools are virtually ineffective if the target population is unable to read the message being transmitted to them. Poor levels of education can also act as an obstacle to effective risk communication. Deprived of basic skills, for example, a person will be unlikely to understand the statistics included in a risk communication or the specific risk factors being explained to him or her.
- **Language** — Language is an obvious obstacle to risk communication, as it is with all communication topics. It would seem that one would need only to learn the language of the target audience to mitigate this issue, but the answer is not always so straightforward. In many communities, there are several languages spoken and several dialects of each language. Throughout the world, there are more than 6,000 recognized

languages. It is a common misconception that all people of one ethnic group or nationality will speak the official language of their particular group or country. Such a misconception can easily lead to risk communication not reaching target audiences. Unfortunately, even if one does learn the target population's language, there may be particular abstractions and colloquialisms that cause common terms and phrases to have vastly different meanings between two speakers of the same language. When such misunderstandings occur in risk communication, the result can be counterproductive at best, and deadly at worst. In the wake of Hurricane Mitch, 1998, the non-profit organization World Vision distributed powdered chlorine to many villages in Central America for purifying water. There was found to be widespread appropriate use of the chlorine in villages where Spanish was the primary language, but inquiries revealed that indigenous villagers (whose first language was not Spanish) were using the chemical to wash their clothes. Because the powdered chlorine was distributed to these indigenous villages without instructions in their native language, they were unable to properly utilize the preventive measure. In this instance, the primary outcome was many gastrointestinal illnesses that probably could have been avoided. Nevertheless, consequences could have been much worse if villagers had ingested the chlorine in deadly doses because of poor risk communication (Swanson, 2000).

- **Access to Technology or the Media** — Risk communicators regularly employ the media to convey a message to a target audience. This is particularly true during sudden-onset disasters, where the media become the primary, if not only, source of communication between emergency responders and the public. Understandably, this form of communication is only possible if the target audience has access to television, radio, newspaper, or the Internet. When access is substandard or nonexistent, risk communicators are presented with a formidable obstacle. The Internet, for instance, which digitally connects the world's population at ever-increasing rates, is a newer communications form that has rising (yet far from blanket) rates of coverage. In addition, while the Internet is an effective risk communications medium, there are certain demographic groups that do not yet enjoy equal or regular

Internet access (e.g., the elderly, the poor, recent immigrants, and transient populations). In poor rural areas, lack of access is evident in older communications technologies as well, including radio, television, and telephone. While they may be the message vehicles of choice for risk communicators, researchers, and the media, these forms of communication may not be effective in all communities and for all audiences. Therefore, to mitigate this obstacle it is important to identify and utilize alternate, non-technical forms of risk communication.

- **Class Structure** — Social scientists have focused considerable research upon the study of community stratification (Cockerham, Lueschen, Kunz, and Spaeth, 1986). *Social stratification* refers to the way certain societies' populations are divided into hierarchical groups based upon inequality. Every society has some form of social stratification, the United States included. The way in which select communities practice this discrimination and oppression effectively limits access for each successive drop in class ranking. These institutionalized hierarchical systems have remained steadfast for decades, if not centuries. Their effect has gone beyond the basic socioeconomic factors of wealth and power, and has created a psychological reality that cannot be underestimated by outsiders who may not fully understand their influence. Risk communicators attempting to transfer their message effectively to populations with such forms of class structure will face a formidable challenge. They will likely need to garner an authoritative grasp of the history and culture of their target audience, much of which may not be formally recorded. It is likely that they will encounter strong resistance unless they can find a diplomatic way to appease all groups while still achieving their stated goals. Furthermore, they will find resistance to change if the risk reduction message requires people to behave unlike one in their caste would (even if that behavior would reduce personal risk).
- **Poverty or the Effects of Poverty** — Poverty, as it influences behavior, access, and opportunity, is an obstacle to risk communicators almost everywhere. Poverty and disasters are intimately connected, as poverty is often the cause of disasters by means of forcing poor populations to live under conditions that directly place them at great risk, and limiting

their access to vulnerability reduction measures. These poor often live in precarious conditions because they have no other viable, easily identifiable alternatives. Sometimes, even when their vulnerability to disasters is extreme, little is done to mitigate hazard risks despite extensive knowledge about them. People living in extreme poverty are often unwilling or unable to participate in conventional risk reduction measures that do not fully account for their poverty. If risk communicators do not consider the economic means and monetary constraints of the people to whom they are communicating the risk, their message will surely fall upon deaf ears. Many poor people live in riskier places because they cannot find alternate housing, not because they do not know the risks. Simply informing them that they are at high risk from landslides or floods, for instance, would do little — even if alternatives are offered — unless those alternatives are viable within their financial bounds.

- **Cultural Understanding** — Incorporating cultural context into risk communication can be very difficult. Risk communicators must fully understand the ways in which not only their words but also their actions, tone of voice, gestures, dress, and approach to discussion, among others, will influence the efficacy of reducing risk. With all population groups, especially those of uniform backgrounds, cultural sensitivity to group-specific attributes must be observed. Not doing so can result in a myriad of negative outcomes, from a communication breakdown to anger, insult, or an increase in risk. However, cultural barriers are not impenetrable if risk communicators avoid ethnocentrisms and utilize creative measures that accommodate local norms.
- **Cultural and Social Norms** — In addition to understanding how the communication message is received and interpreted by a target audience, communicators must also understand how their proposed solutions mesh with the audience's culture and way of life. Oftentimes, behaviors that contribute to risk are not performed out of ignorance of their negative impact, but rather because they serve some long-standing cultural significance that is not easily abandoned. Communicators must understand the origin of beliefs and behaviors in order to devise solutions that accommodate them.

- **Lack of Community Sponsorship** — Effective risk communication is undoubtedly enhanced by the official support of a local community and its government, even if that support is marginal. This enhancement is pronounced if there is great trust in the government or community organizations and officials that are serving as champion to the cause being communicated. Agencies and organizations can show their support by performing actions ranging from making official statements of endorsement to passing laws requiring or prohibiting certain activities. However, if the community or its government does not support the public information message, that stance becomes a severe detriment to communicators.
- **Denial or Apathy** — Denial is common, but it can be overcome. People tend to construct psychological barriers to preparedness, either minimizing the likelihood of a threat by assuming it will not affect them if the threat becomes a reality, or feeling that there is little they can do to mitigate the consequences of a disaster. By concentrating on the survivability of people in the event that realistic hazards do occur, preparedness education emphasizes positive steps for positive outcomes. The assumption is that people can learn to prepare, and those who do will cope with the disaster better. Apathy differs from denial in that individuals simply lack the motivation to change their behavior despite knowledge of its risk reduction potential. Complex psychological and social processes, in addition to a simple unwillingness or true lack of care, can cause apathy.
- **Risk Perception and Heuristics** — In order to manage a risk, one must also be able to judge the relative seriousness of that hazard in comparison to other hazards. Risk analysis is what disaster managers use to compare and rank community hazards. For laypeople, however, in the absence of such technical and involved analysis, the mechanisms by which they *perceive* the hazards that threaten them can be very different and very complex. Risk perception is a field of study that attempts to explain why people fear the things they do (and why they do not fear other things). Risk perceptions more often than not differ significantly from actual statistical risks, causing people to fear most things that are statistically unlikely to harm or kill them, while risks that are more probable are of

no concern. *Heuristics,* which are types of thought processes that influence the ways people perceive and analyze information and experiences (oftentimes called rules of thumb), have a similar effect on communication efforts. Several heuristics must be considered. For instance, once people make an initial judgment (i.e., first impression), they believe with overwhelming confidence that their beliefs are correct. This phenomenon, called the *overconfidence heuristic,* is the result of people being unaware of how little they know about a risk, and how much more information they need to make an informed decision. More often than not, people believe that they know much more about risks than they actually do. Risk perception and heuristics are described in detail in Chapter 3.

Risk communication experts Vincent Covello and Peter M. Sandman state that one of the greatest risk communication obstacles comes from the uncertainty, complexity, and incompleteness of statistical data associated with disasters. The uncertainty component of natural and technological hazards is what distinguishes the issues managed by emergency management communicators from those of the public health community (who work with risks for which there exists much greater understanding of associated likelihood or consequence). To make effective mitigation or preparedness decisions, emergency managers rely upon risk assessments that have far fewer data points, and they must rely upon a significant amount of expert judgment (over clinical experience, for example). Covello and Sandman contend that such a lack of more concrete data presents a critical challenge, primarily because the communicators may only be able to state that the risk in question falls "somewhere between serious and nonexistent" (e.g., earthquake risk).

Influential External Variables

In addition to obstacles that get in the way of communication efforts, there is another set of factors that influences how messages are received and perceived by recipients. These factors exert pressure and interference on one or more of the stakeholders in the campaign (including both the communicators and the recipients). Unfortunately, communication cannot be performed within the confines of a controlled laboratory where all factors are restricted in a manner that maximizes the transfer of messages from communicator to recipient. In the real world, there are competing interests, a multitude of voices, and a full gamut of individual agendas.

External variables can exert both positive and negative pressures on a communication campaign. The key is being aware of the variable and planning for it as best as possible either to capitalize on the benefits or to minimize the impediments. Negative influence most often comes in the form of information or belief that opposes or contradicts a key campaign message. This form of outside influence can originate from a multitude of sources, including private interests (businesses), cultural behaviors or histories (including customs and traditional behavior), popular attitudes and beliefs, and many others. The more a planning team is able to identify these factors and adjust for their presence accordingly in their messages, channels, and methods, the less of a detrimental effect they will have on the outcome of their efforts. Each will represent a conflict between the communicator's message and the will of the audience to comply. By directly addressing the concerns or misperceptions of the audience that pertain to these factors through the content of the message, the influence of many if not most of these negative variables can be reduced or avoided.

Outside variables may also have a positive effect on the outcome of the campaign, especially if they are properly identified and appropriately utilized. For instance, the occurrence of a disaster anywhere in the world can have the positive effect of raising awareness both inside and away from the disaster area. During these so-called windows of opportunity, the public has been found to be more apt to pay attention to disaster preparedness messages, even if they were not affected. How much they actually change their behavior will rest upon the abilities of the campaign team to seize upon the increased salience and get their point across effectively. By latching on to the positive or parallel nature of these factors, the campaign may stand to gain considerably.

For both positive and negative external variables, it is key that identification be made as early in the planning process as possible. The difficulty in achieving such early recognition is a matter of access to the information sources by which such discovery is possible. This could be as simple as performing a basic Internet news search or interviewing members of a certain population. The following list of categories of external influential factors is representative of what may be encountered, and is by no means exhaustive.

- Current events
- History
- Timing
- Media interest
- Public policy
- The three agendas (government, public, media)
- Competing messages
- Popular culture
- The economy
- Audience characteristics (described in Chapter 3)

Sidebar 4.2: Parental Instinct Presents an Unforeseen Obstacle to Evacuation Preparedness and Planning

A survey conducted in the summer of 2008 found that 63% of parents faced with an evacuation order would disregard those instructions if their children were in school at the time. These parents indicated that rather than flee the area in danger, they would immediately drive to their children's schools to pick them up themselves. Such behavior would likely further hinder both general evacuation efforts by causing more traffic, and the organized evacuation efforts at the individual schools. The survey, which was commissioned by the Columbia University Mailman School of Public Health and the Children's Health Fund, found that nearly 45% of parents did not know to where their children would be evacuated by the schools, which could help explain a certain amount of the anticipated resistance. However, emotional factors would also play a strong role, as indicated by a mother of three who was quoted by the Associated Press as stating that, "As a mom, you wouldn't be able to keep me away from picking up my children. My first instinct would be to get them at all costs. I would literally run the entire distance to get them. I believe most parents would feel the same." It is important and even more revealing to note that this individual is also the author of a disaster preparedness book entitled "Your Home Office Recovery Plan," a fact that emphasizes the strength of the emotional obstacle.

Source: Matthews, Karen. 2008. Survey Finds Holes in U.S. Disaster Preparedness. The Associated Press. September 12.

SELECTING APPROPRIATE SETTINGS, CHANNELS, AND METHODS

Public education communicators must design their message within a framework that accommodates the ways in which their target audiences acquire information in general and specifically as it pertains to hazards and risks. Each audience will be unique in this regard as a feature of their means, culture, social networks, and other factors as described in Chapter 3. For instance, people living in poverty or who are marginalized in society oftentimes gather information to make decisions through informal social networks rather than from newspapers, official government sources, or other formal communications methods. Members of these groups may mistrust government and other "official" information sources, leading them to ignore or disregard their messages. Another group, transient populations (e.g., tourists and business travelers), tends to be limited in both time and motivation when learning about particular hazard risks simply because they are outside their normal social and physical environment (and therefore less likely to perceive a hazard risk that might not be readily apparent). Special needs populations, which include the elderly, young children, the disabled, and the illiterate, for example, must each be approached in a manner that addresses their particular method of perception and learning. In defining the audience, and through market research, these preferences will have been identified.

Three factors must be considered when determining how audiences will be reached and how messages will be delivered to them. They are

1. Settings
2. Channels
3. Methods

Settings

Settings are the first of three factors that communicators use to frame their preparedness message. As the name suggests, settings are the situations in which communication occurs (see Figure 4.1). It has been found that learning is not uniform in all situations, and that while certain settings can hinder the communication process, others can be highly effective at fostering the reception and application of a message. It is not hard to imagine the difference in success of learning for students in the classroom as compared to in an arcade, for instance. However, in reality, the differences are often

FIGURE 4.1 Preparedness campaign billboard in western New York uses a fear appeal to urge commuters to visit the campaign Website to gain information about emergency preparedness.

not so obvious at first. Identifying what settings are most appropriate for an audience is therefore key to increasing the success of the campaign.

There are three primary factors that distinguish a setting:

1. Time
   - Time of day (e.g., morning, afternoon, evening)
   - Day of the week (e.g., weekday, weekend, Sunday)
   - Time of year (e.g., hurricane season, during summer vacation, wildfire season, early winter)
2. Location (e.g., at home, at work, in school, while on vacation, in a car or other form of transportation, at the doctor's office, in a store, at a community center)
3. Situation (e.g., while having dinner, while gathering with friends or peers, while shopping, while attending a school assembly, while visiting a county fair)

The occurrence of an actual disaster is often viewed as the ideal setting within which the topic of disaster preparedness may be lifted to the forefront of both the public and the media agendas. However, various non-disaster times also occur throughout the year when public interest in the topic of preparedness may be raised to a comparable degree. During these key times, the news media recognize the salience of the issue and are therefore more likely to pay attention to events, use material from

FIGURE 4.2 Oakland, California, September 18, 2007 — As part of National Preparedness Month, FEMA employees manned information booths to help build public awareness and provide booth visitors with preparedness materials for home, family, school, businesses, and individuals with disabilities and special needs. (*Source*: FEMA News Photo.)

press releases, request interviews from key officials in the organization, and attend preparedness-related events. A campaign kickoff event, in which a key official announces the start of a preparedness or other effort at a planned ceremony, can be very effective at drawing positive media attention. National Preparedness Month is probably the single greatest time of the year for an organization to hold a disaster-related kickoff event (see Figure 4.2). Of course, any time of the year is appropriate, as long as there is a campaign, event, or other news to announce. Disaster seasons allow another opportunity to garner public and media attention. While not all hazards have designated times of year when their likelihood goes up, many do. For example, a set hurricane season runs each year from June 1 to November 30. The media grasp upon the opening of this season, which provides an excellent opportunity to draw their attention to your event. Other examples of hazards with seasons include wildfires and severe winter weather. Anniversaries of major disasters, especially in the first years following that disaster, can draw equal (if not more) attention as disaster seasons. Eight years after the September 11 attacks in New York and Washington, there is a marked increase in media attention to terrorism during the weeks surrounding that important date. Late August has drawn attention to hurricane preparedness every year since Hurricane Katrina. The flurry of preparedness activity surrounding the 100th anniversary of the Great 1906 San Francisco Earthquake proves that events do not need to be recent. Finally, there is the ever-present availability of "piggybacking" activities. Piggybacking is achieved by injecting a preparedness message into another important event or activity that is salient and therefore in the media spotlight.

The key to selecting a setting is to determine not only where and when you can best reach the target audience, but also where and when they are most likely to be attentive to, receive, understand, and act upon your message. The best places are those that help to establish message credibility. Unfortunately, it is often the case that times, places, or situations that are ideal for reaching the audience are not ideal for delivering the message. For this reason, we must also consider communication channels.

Channels

A channel is a route or mechanism by which a message is delivered. Within each broad channel category, communicators can reach their target audience through many different individual sub-channels. The five primary message channels, and examples of sub-channels within each category, are listed in the following:

1. **Interpersonal Channels** — Communication methods included in these channels are those that rely upon the direct personal interaction between a communicator and an audience member. Oftentimes, the communicators are people who are highly trusted by the audience member, who serve as authority figures, or who are in some way influential to the audience member. As such, interpersonal communication channels tend to be the most effective at motivating people to change. The human interaction helps to mobilize people, and can provide social support to maintain the behavior changes that are advocated. However, because of their reliance on high-effort one-to-one interactions, they do not necessarily have the immediate and broad reach of other channels such as, for instance, the mass media.

2. **Group Channels** — Communication methods included in these channels are those that create or exploit gatherings of target audience members, whether to discuss the intended message or for another reason entirely. Like interpersonal channels, group channels can help communicators to enjoy some of the trust and influence that exist within the group, but by talking with several individuals at once, the message reach is much greater. Additionally, by creating discussions among peers with similar wants and needs, many doubts can be alleviated not only by the communicator, but also by peers themselves (thereby lending more credibility to the message). However, these methods are also similar to interpersonal methods of communication in that they require significant effort to reach larger audiences over time.
3. **Organizational and Community Channels** — Communication methods included in these channels include those that rely upon official, established entities operating within the community that are able to interact with members of the target audience. These are many of the same organizations and entities that were described in Chapter 3. Businesses, NGOs, local and state government organizations, religious organizations, and many others can help you to communicate your message to the intended audience. While they may not be as willing as those individuals and groups described in the interpersonal and group channels to have at-length discussions about emergency preparedness, they may help communicate your message by posting signs, distributing fliers, playing videos on showroom monitors, making announcements, or referring audience members directly to you or your Website. These organizations, groups, and other entities often operate on a much larger scale than do the individuals or groups mentioned in the previous two channels, and can therefore have a much greater overall message reach. In addition, when they are recognized within the community, and enjoy a high level of respect among community members, they can be very effective conduits, regardless of whether their mission relates to emergency preparedness. Their involvement in your campaign is, in essence, a sponsorship of your message.
4. **Mass Media Channels** — Communication methods included in these channels include those that broadcast a message out to large audiences, in most instances in a unidirectional manner. Mass media campaigns have been tried and tested in the public health field, and have already found success in the emergency

management field with emergency preparedness messages. The news media, which include television, radio, print, and Internet news outlets, are the most recognizable component of the mass media. However, the mass media is a much broader category that spans many other conduits, including:

- Magazines
- Billboards
- Theater
- Trade publications
- Cinema and television shows
- Talk shows
- Advertising
- Direct mail systems
- Entertainment programming

5. **Interactive Digital Media Channels** — Interactive digital media channels represent the most recent addition to the short list of communication channel categories. Communications methods included in these channels include those that exploit the Internet and other digital communication media to allow for direct interaction between communicator and recipient. While this communication form shows incredible promise in terms of its expanding communicators' reach and success, it still carries inherent problems (see Sidebar 4.3). Sub-channels found within this category allow for greater matching of message to intended audience; but unlike the regular mass media channels, these allow for audience feedback. The digital media are already recognized as one of the, if not *the*, fastest growing communication tools. With proper Webpage development and design, organizations can quickly and effectively reach large numbers of people with only a small investment in time and resources. Websites also allow for an immediate two-way interaction that is more difficult to attain through banners, flyers, and other message dissemination methods. A Website can be designed to simply communicate the developed message by including descriptions of planning initiatives with upcoming meeting dates, times, and minutes from meetings, or it can be highly complex and interactive with links to mitigation and hazard resources and sites and opportunities for questions and answers. Websites are also good venues for posting questionnaires for citizens to determine their perceptions of hazards and risks in the community or state, as well as provide an additional

outlet to generate feedback on issues. In many ways, this channel is a fusion of the previous four channel types.

Sidebar 4.3: Problems Associated with Interactive Digital Media Channels

Credibility: Most (but not all) Internet users are aware that anyone with Internet access can post information regardless of its accuracy. Communicators must therefore be able to demonstrate a high level of credibility if their message is to be taken at face value by target audience recipients — a requirement if they hope for their message to be believed. *Healthy People 2010,* a public health communication effort, recommends that communicators disclose the following information in order to build credibility among recipients:

- The identity of the developers and sponsors of the site, how to contact them, and information about any potential conflicts of interest or biases
- The explicit purpose of the site, including any commercial purposes and advertising
- The original sources of the content on the site
- How the privacy and confidentiality of any personal information collected from users is protected
- How the site is evaluated
- How content is updated

Recipient Access: The average computer user is affluent and well educated. Although Internet access is increasing, it is not universal like that of television and radio. First, there is the issue of physical access. While it is estimated that 98 to 99% of American homes have at least one television, only 75% have Internet access. Second, there is the question of technical proficiency. Unlike passive learning as occurs on television, the Internet requires not only an active learning style, but also an understanding about how to use the hardware and software required to find and access preparedness information. The gap between those with and those without Internet access is called the *digital divide,* and tends to affect those living in poverty, the elderly, and other special needs populations.

(Adapted from the National Institutes of Health.)

Methods

The actual transfer of message content from communicator to recipient is performed through one or more communication methods. A communication method is an actual item, action, interface, or event that communicators use to draw the attention of the recipient, and to inform him or her of the behavior change that is necessary (or how to access that information). As was true with the settings and the channels selecting, communication methods must fall within the access, interests, and learning styles of the recipient audience members. Several communication methods are specific to communication channels, including the following:

- Methods unique to the interpersonal channel include discussions with:
  - A teacher or professor
  - A friend
  - Family members
  - An insurance agent or financial planner
  - A building contractor
  - A coach or counselor
  - A religious figure (e.g., priest, pastor, rabbi, imam)
  - A doctor
- Methods unique to the group channel include:
  - Classroom discussions
  - Adult education
  - Neighborhood association meetings
  - Parent–teacher association meetings
  - Municipal and civic club meetings
  - Town hall meetings
  - Office meetings (including brown bag lunches, for example)
  - Religious gatherings
  - Volunteer events
- Methods unique to organizational and community channels include:
  - Signs or free advertising in grocery stores, home improvement stores, hardware stores, or other businesses
  - Information included in local utility bills or phone books
  - Signs in subways, buses, or other forms of public transportation
  - Signs or brochures offered in waiting rooms
  - Information in hotel rooms or on tourist maps

  - Message delivery by community advocacy groups
  - "Bag stuffers" placed with purchases in stores
  - Information printed on packaging (e.g., on a milk carton or cereal box)
- Methods unique to the mass media channel include:
  - Live broadcasts at preparedness events, or recorded interviews
  - Feature stories on preparedness events
  - Mentions in news stories or programs
  - Preparedness subject matter incorporated into TV shows or other television programs
  - PSAs
  - Magazine advertisements, articles, or messages
  - Newspaper editorials
  - Columns in newspapers and magazines
  - Press releases
  - Press kits
  - Press conferences
  - Call-in shows
  - On-air announcements
- Methods unique to the interactive digital media channel include:
  - Emergency preparedness Websites
  - Preparedness "Webinars" (online seminars)
  - Online bulletin boards
  - Online courses
  - Online presentations
  - Online videos
  - Online forums
  - Newsgroups
  - Instant messaging services
  - Email preparedness messages
  - SMS (telephone messaging) messages
  - Digital video conferencing (DVC)
  - Web-based surveys
  - Online preparedness games and activities
  - Blogs
  - Links from other Internet sites to your Website, or links to preparedness information
  - Informational kiosks

There are many other methods, which can be used on their own or combined, that apply to more than one communication channel. These are limited only by the creativity of the planning team, and include (among many others):

- **Brochures, fliers, fact sheets, newsletters, and inserts** — Brochures, fliers, fact sheets, and newsletters are relatively inexpensive to produce and can be useful in reaching a wide audience. The materials can be created either internally or externally, and may be something that a business with printing capacity could donate to the campaign in-kind. If created electronically, these materials can be generated and distributed for little or no cost. As with all publications, all materials should be reviewed for accuracy and message, approved by appropriate staff, and tested with members of the target audience before they are distributed. The materials should be clear and easy to read and understand. The message should include a designated department or contact name and phone number in case a reader is interested in learning more. One way to quickly reach a wide audience is to work with a partner, such as a utility, to include the materials as an insert in a bill. Alternatively, partners may agree to distribute the materials to their clients or customers. Examples include electricity or phone bills, the Yellow Pages or phone book, at grocery or department stores, in government buildings, and at libraries.
- **Festivals, fairs, and other public events** — Public events provide unique opportunities for public educators to reach large numbers of their intended audience in a relaxed and informal atmosphere. The events may be initiated and managed by the organization for the purposes of communicating with the public, or the organization may simply decide to participate in an event being managed by another organization for a related or unrelated purpose (such as would be the case with an annual county fair). The communication opportunities at a large public event are diverse, and may range from an informational booth to performances and demonstrations. Oftentimes, even if the event organizers are charging a set fee for booth space or other involvement, it is possible to garner donated time or space because of the public-service nature of the communication effort. These events allow for the use of several of the different available communication channels. Additionally,

they allow organizations conducting public education efforts to talk to their audience about their experience with hazards and try to use this information to better tailor their education efforts in the future. A particularly beneficial aspect of this form of communication is that the event provides people with an opportunity to ask questions face-to-face.

- **Advertisements** — Many organizations of all types are often willing to donate or otherwise provide advertising and other communications means to disaster preparedness programs. Businesses can be tapped to assist in educational efforts both within and outside of the realm of their offices. Internally, efforts could include employee preparedness campaigns or materials distributed to employees to bring home to share with friends and family. Externally, businesses may help by providing advertising time or space, materials and equipment, endorsement of the campaign, inserts in bills or newsletters, mailings, giveaway items for events, and many other things as appropriate for each business type. Religious groups and churches can distribute pamphlets or have speakers at their many events, and may even sponsor special events focusing on disaster preparedness topics. Social organizations operate in a very similar fashion, and would likewise have the capacity to foster public education of its members or the public in general. Using guest speakers and experts, public education teams can reach these captive or interested audiences.
- **Formal education (in schools)** — Schools are natural places of learning, and as such they can foster effective disaster-specific public education among students. Schools are very often receptive to public education efforts aimed at decreasing the vulnerability of students. In-school programs often involve disaster safety mascots (such as Sparky the firedog), representatives from the emergency services (a firefighter or police officer), or other public officials. In school, education is either classroom-based or performed in large auditoria. Some schools may allow posters to be hung on classroom and hallway walls, or to have representatives from the organization present at large events where a table or other educational venue can be set up. Educational materials that target student populations could include special disaster-specific courses (long or short), distributed materials and fact

FIGURE 4.3 Emmitsburg, Maryland, March 10, 2003 — FEMA's National Emergency Training Center is the site for dozens of classes, including sessions that train CERT leaders from across the country. (Photo by Jocelyn Augustino/ FEMA News Photo.)

sheets, information integrated into regular coursework, games, coloring books, and contests, among others.

- **Adult and other "out-of-class" education** — Special courses can be designed to teach adults or out-of-class students about disaster preparedness in much greater detail than is possible in distributed documents or media announcements. Programs such as CERT have been widely successful in creating local cadres of citizens prepared to respond to disaster events. Organizations could create CERT-type courses that are tailored to the specific needs and abilities of their audience (see Figure 4.3).
- DVDs and books
- **Emergency drills and exercises** — Emergency drills and exercises test one or more aspects of disaster response by exposing participants to an imagined or recreated disaster scenario. Tabletop exercises involve the reading of a disaster narrative followed by questions and discussions. Functional and full-scale exercises involve action that is more realistic, oftentimes with

actors playing the various roles involved in an actual response. Drills and exercises are effective on several levels, including their ability to raise awareness to a disaster, their ability to highlight deficiencies in preparedness, and the reinforcement of effective practices that save lives and property. Additionally, they tend to garner media attention. The greatest detriments to this communication method are the cost and planning difficulties it involves.

- **Alternative media** — For most audiences, and most common messages, television, radio and print are the conventional methods of delivery. However, special media is required for some special needs audiences, such as Braille for the blind.

Other examples of transmission methods include:

- Posters
- Bookmarks
- Handbooks or manuals
- Checklists, to-do lists, and shopping lists
- Disaster preparedness–themed game boards
- Cartoons and comic books
- Photographs
- Theatrical performances
- Informational pages in telephone books or other directories
- Information printed inside matchbox covers
- Disaster information displays or kiosks
- Markers or signposts established to remind residents of past disaster incidents
- Tours containing hazard and preparedness information
- The creation of a speaker's bureau (to guarantee that any service organization, parent–teacher group, or church group can learn about emergency planning)
- Disaster preparedness camps or workshops
- Pre-assembled do-it-yourself preparedness kits
- Public service announcements
- Exhibits
- Conferences
- Disaster simulations
- Disaster awareness events
- Branded merchandise
- Communication programs internal to organizations (such as with employees, members, parishioners, etc.)

SELECTING COMMUNICATION CHANNELS AND METHODS

Communication channels and methods are selected according to how the audience is most likely to receive the information being communicated, and the resources and abilities of the program. Communicators can use a single channel, or even a single method, but they are more likely to see a desired change in target audience behavior if they employ a combination of channels and methods to broadcast their message. In addition, when resources allow, the format of the message can be varied and communicated to the audience in many different ways, as often as is possible and practicable. The generally accepted belief is that more channels and more methods translate to wider exposure.

To maximize exposure, planners should focus on channels with which the target audience has regular contact, and that they perceive as credible. Familiarity with the channel and with the communicators themselves helps to build message acceptance. For instance, projects that use popular entertainment media (e.g., movies, soap operas, radio plays, music, theater, comics, and so on) have been found to be particularly successful because target audience members can identify with the heroine or hero, or a well-known idol, giving them a motivational push in the desired direction of change.

Because the impact of a single interaction can be short-lived, repetition of the message helps to ensure that learning occurs. Static images and messages, such as those on posters or PSAs, eventually become monotonous, and target audiences may stop paying attention to them completely. For this reason, it may be necessary to change channels, materials, and messages from time to time.

Interpersonal contact is indispensable to a public education effort, regardless of the channel or combination of channels selected. The mass media is adept at building interest in a topic, but personal consultation or motivation by a trusted caregiver can make all the difference between mere awareness of a topic, having a positive attitude toward it, and actually adopting the new behavior. Personal communication can strengthen the lessons learned through all other channels, and can fill in the gaps where mass media and other less personal channels fail to penetrate. It is always wise to include some avenue that allows for interpersonal contact between the target audience and communicators.

For each channel and method under consideration, communicators should measure the pros and cons. This can be done by considering the following questions:

- What access does your organization have to the channel or method?
- How many people will be exposed to the message transmitted by the channel or method?
- Will target audience members pay attention to the method transmitted by the channel?
- Does the intended audience accept and trust the channel or method?
- Can the target audience be influenced by the channel or method?
- Is the channel or method appropriate for conveying information at the desired level of simplicity or complexity?
- If skills need to be modeled, can the channel or method be used to model and demonstrate specific behaviors?
- Can the channel or method provide interactivity between the target audience and communicators?
- Using the channel or method, how long will the message last (i.e., will it be transmitted once, lasting seconds, or will it last for weeks or months as with signs or print materials)?
- Can the channel or method allow the intended audience to control the pace of information delivery?
- Can the target audience control the pace at which information is delivered?
- Are there any opportunities for repetition with the channel or method?
- Does the channel or method increase the likelihood that the target audience members will retain the message?
- Does your program have the resources to use the channel or method?
- Is the channel or method appropriate for the activity or material you plan to produce (see Figure 4.4)?
- Will the channel or method reinforce messages and activities you will develop through other routes (in order to increase overall exposure among the intended audiences)?

Once channels are identified, the basic principles of advertisement should be observed. The following formula has been time-tested for use in determining the likelihood of success when exposing a message to the public:

$$\text{Reach} \times \text{Frequency} = \text{Success}$$

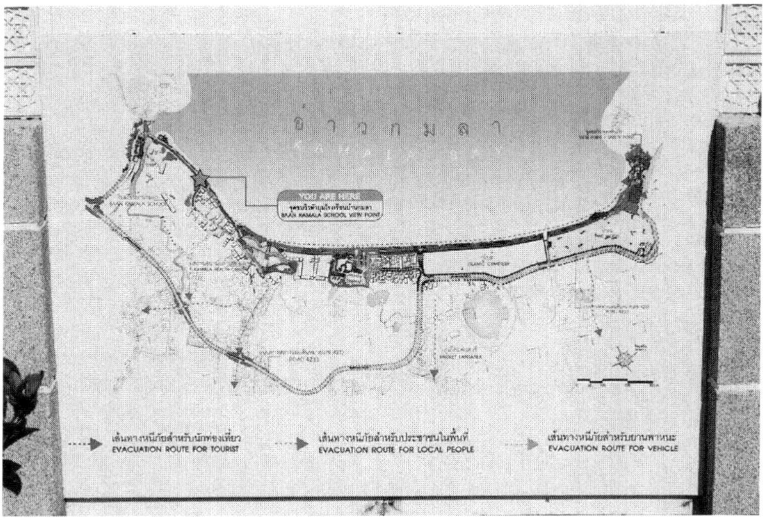

FIGURE 4.4 Tsunami evacuation sign, Phuket, Thailand. This evacuation sign is intended to provide a preparedness message to an audience that generally has a very short attention span to such matters — tourists on vacation (Coppola, 2008).

where Reach is the percent of the target population in a geographically defined area exposed at least once to the message during a specific time frame, and Frequency is the number of times a message has been broadcast to a target audience.

A Canadian public health communication program that promotes healthy living, Active2010, has identified the following strengths and weaknesses of several of the methods listed previously:

Strengths	Weaknesses
Print Media (newspapers, journals, and magazines)	
Large reach Can be free or low cost Information can be kept and shared Information covered more thoroughly than other mass media channels Works well with complex messages	Literacy implications Message must be newsworthy Exposure can be limited to one day Possibly low emotional appeal Can be expensive

Strengths	Weaknesses
Newsletters	
Reaches opinion leaders Low cost High message preservation Responsive	Labor intensive Tends to preach to the converted Literacy implications
Radio	
Large reach Range of formats Can be low cost or free Can be interactive with call-ins Use of scripts can give the impression that messages are live and from the radio hosts Timely and repetitive Possible use of celebrities Literacy not an issue	Can be expensive No visuals Little or no placement control PSAs tend to be played during off-hours May only reach a specific audience Limited to radio reception area Limited audience retention
Television	
Extensive reach Can be free Different target groups reached Visual impact	Can be very expensive High level of complexity is possible May be a limited reach PSAs run infrequently and at low viewing times Messages may be obscured by commercial clutter Retention can be difficult
Posters, Signs, and Billboards	
Attract attention Wide reach Captive audience High message repetition Geographically focused Visuals can have great impact	Can be expensive Low audience specificity Limited to short, simple messages
Telephone	
Confidential, personal, and private Interactive Follow-up is possible	Cost can be high Intrusive Labor intensive

Strengths	Weaknesses
Telephone (continued)	
Efforts can be directed to specific audiences Can be inexpensive	
Electronic (including CD-ROM or DVD)	
Large reach (but select audience) Interactivity is possible More youth-friendly Easily updated	Literacy barrier High cost for equipment Requires skills and training
Internet	
Can reach large numbers of people quickly Can instantaneously update and disseminate information Control over information is possible Information can be tailored to specific audiences Can be interactive Information can be presented in a graphically appealing way Can combine the audio/visual benefits of TV or radio with the self-paced benefits of print media	Can be expensive Many target audiences do not have Internet access Audience must be proactive (they must find and access the Internet themselves) Newsgroups and chat rooms may require monitoring Can require maintenance over time
Mail	
Reaches specific area (e.g., zip code) Information can be retained Follow-up is possible Appeals to visual learners Can be directed (named individually)	Cost can be high Can get lost or seen as junk mail Literacy can be an issue Can create negative associations
Point of purchase (such as in a hardware store)	
Timely Immediate reinforcement Can be interactive (e.g., demonstrations) Information is targeted where it is needed Good opportunities for partnerships	Relatively small reach Loses effectiveness over time Can discriminate against low-income people Difficult to partner if controversial Depends on others to get the message out

Strengths	**Weaknesses**
Presentations	
Interactive Specific and captive audience Timely information Control of content	Relatively small reach Can be costly in time and resources Poor retention Presenter's beliefs and biases may affect message
Training	
Reaches specific audience Strong multiplicative power Builds skills Peer-to-peer support Interactive	Low motivation if attendance is not voluntary Different learning styles may not be accommodated Limited one-on-one consultation Commitment to follow-through not guaranteed
Informal networks	
Interactive Comfort of cultural similarities Familiar and safe Provides access to other networks Social support provided	Information may be biased or unreliable Focus on experience may be narrow Requires a certain personality Can be exclusive to some Limited, homogenous group
Contests, Fairs, Fundraisers, and Festivals	
Large reach Interactive and fun High visibility and interest Opportunity for media coverage Captive audience	Difficult to follow up on and evaluate Difficult to tailor to a specific audience Labor- and resource-intensive Short lifespan and narrow focus
Interpersonal Channels	
Most effective for teaching and helping More credible Permits two-way discussion Can be motivational, influential, and supportive	Can be expensive and time consuming Can have limited target audience reach Can be difficult to access these channels Sources need to be convinced and taught about the message themselves

Strengths	Weaknesses
Organizational and Community Channels	
May be familiar, trusted, and influential May provide more motivation/support than media alone Can be inexpensive Can offer shared experiences Can reach larger intended audience in one place	Can be costly and time consuming to establish May not provide personalized attention Organizational constraints may require message approval May lose control of the message if adapted to fit organizational need

SELECTING COMMUNICATORS

For many (although not all) of the communication channels and methods listed previously, the planning team must select and train communicators to transmit their messages to the intended audience. The task of selecting communicators is one that deserves careful consideration of the planning team because the communicator becomes as important as the message itself. Risk communication messages are so diverse that almost everyone, including children, parents, the elderly, educators, the illiterate, and employers, are capable of communicating them. It is the variance between and particulars of message settings, channels, and target audience members that determine which of these various communicators is likely to be the most effective at bringing about actual behavioral change. These different communicator characteristics must be identified and exploited in order to increase the likelihood of message reception *and* acceptance among different target audiences. While some communicators may be great public speakers and have the ability to sway the opinions of large audiences, others may be similarly effective working with individuals through interpersonal interactions. The goal is to make target audience members feel personally addressed and, likewise, taken seriously by the communicator. When assessing and selecting communicators, the following measures may be used:

1. **Speaking ability**: Not all people can communicate a message in an effective manner in such a way as to convince the recipient of their message. One of the most common fears, in fact, is that of public speaking. Others may fail to present arguments coherently, losing the point of their message in transmission. Speaking ability, whether

in one-on-one situations or to large audiences, is central to communication and is easy to assess by looking at experience and successes.

2. **Reputation among audience members (trustworthiness and credibility)**: Message recipients will judge a message not only by its words, but also according to any preconceptions they have about the communicator. The changes in behavior that audience members are expected to make are changes that will likely affect their thinking, their finances, and possibly their way of life. Understandably, people will only be willing to take the prescribed advice if they hold a great amount of trust in the communicator — and for each communicator, trust and credibility will differ according to the audience. Clearly, anyone who has previously betrayed the trust of their audience, or has even been accused of doing such, would have difficulty finding receptive ears. Credibility, or the ability to be convincing, is related to trust and varies by audience for each communicator. For instance, while citizens with legal residency status may find a high-ranking government official to be credible, illegal immigrants who are more likely to be evasive of the government will likely disregard the message entirely.
3. **Subject matter knowledge**: Communicators must have a strong understanding of the topic, or at least of the information contained in the message they are communicating. If they appear to waver at all in their understanding, or if they make a factual error, the audience will negate their message entirely. Even one small factual error, if discovered by the audience, will likely result in complete message failure.
4. **Image of authority**: Audience members will listen to a communicator's message only when they believe that the communicator is more knowledgeable about the topic than they are. For this reason, the communicator's credentials become as important as the knowledge he or she is imparting. With children, for instance, almost any adult will seem convincing. For an educated adult, however, information about the communicator's background will play into the believability of the message.
5. **Obvious lack of vested interest**: Communicators must convince their audience that they are providing risk reduction information simply out of care or concern. If the audience feels that the communicator has anything personal to gain from changes in audience behavior (such as a financial gain resulting from the sale of

some product like pre-assembled emergency kits), or that his or her intentions are politically self-serving, then they will not be likely to change their ways.

6. **Ability to connect, sympathize, or empathize with the audience**: Finally, the audience must be able to relate to the communicator. The communicator can often achieve this through the exploitation of personality, common traits, or characteristics shared with the audience, or through relating common experiences.

No communicator will suit every audience's needs. Therefore, planners will have to have an open mind when considering the ideal candidates. The following categories of communicators illustrate the wide range of possibilities available to the planning team:

- Peers (friends, people of the same age or gender, people of the same financial status, people of the same demographic, etc.)
- Family members (parents, children, brothers, sisters, etc.)
- Teachers
- Employers (managers, business owners)
- Caregivers
- Advocates (union leaders, representatives)
- Retailers (local or national)
- Authority figures (community leaders, elected officials, police officers, firefighters, EMS technicians, etc.)
- Religious leaders
- Community groups (Boy Scouts/Girl Scouts, 4H, Jaycees, etc.)
- Celebrities (athletes, musicians, actors; see Sidebar 4.4)
- Cartoons, puppets, or mascots
- Actual disaster victims

Sidebar 4.4: Smokey Bear

One of the most effective emergency preparedness communicators of all time, and clearly one of the most recognized in the United States, is Smokey Bear. The Smokey Bear campaign was created by the Ad Council in 1944 to serve as the face of forest fire prevention in the United States, educating and urging citizens to take the necessary precautions to both prevent and respond to forest fire emergencies. Over the years, in what has become the longest running public service campaign, Smokey has been depicted in cartoons, animations, posters, radio ads, books, as a costume-wearing mascot, as stuffed animals, and much more. Smokey's message, "Only you can prevent forest fires," is credited with decreasing the acreage of land lost annually to forest fires from 22 million to 4 million. As a testament to the campaign's reach, a 1989 survey found that 71% of American children ranging in age from 5 to 13 recognized the image of Smokey Bear and could recount his fire prevention message (Fuller, 1991). The U.S. Forest Service, the Ad Council, and the National Association of State Foresters administer Smokey Bear, and his image is protected by federal law. Communicators can arrange through their state forester to have Smokey Bear appear at certain events, given that they are large and open to the public.

DESIGN AND DEVELOP MESSAGE CONTENT

Upon completing the formative research behind the campaign, planners can begin to develop messages that are based on their findings. By ensuring that they consider the basic tenets of communication science, practitioners can maximize their likelihood of message effectiveness. This section provides an overview of these considerations.

Message appeal is one of the first and most critical decisions communicators make when designing their messages. Three of the most common appeals used in persuasion attempts are humor, guilt, and fear. Each is described in the following:

- **Humor**: Humor appeals have enjoyed modest success in motivating people to engage in particular behaviors in the past. Generally, when people find themselves entertained by a humorous message, they are unable to carefully process the most important points that the message is trying to transmit. Because of these inherent problems, humorous appeals work best in simple campaigns such as advertising because they are able to increase brand recognition without having to make a compelling case for product use. Brand recognition is not guaranteed with humor appeals, however, because oftentimes people become so wrapped up in enjoying the joke that they fail to pay attention to the brand or behavior being promoted by the message.
- **Guilt**: Messages that use guilt appeals are relatively easy to design because it is rather easy invoke feelings of guilt from people who

FIGURE 4.5 Public health education has long used fear appeals to encourage smokers to quit their habit. This series of anti-smoking packaging in Singapore uses an extreme form of fear appeal (Coppola, 2007).

have failed to engage in a particular behavior. Guilt appeals often grasp upon message recipients' care and concern for loved ones. Research has found guilt appeals to be highly effective when messages are exchanged in face-to-face encounters. However, target audience members do tend to downplay the importance of, and even avoid, people who make them feel guilty. For this reason, guilt appeals are not recommended for campaigns that are mostly using mass media to spread the message.

- **Fear**: Fear appeals, which convince target audience members that something bad will happen to them or people close to them if they fail to engage in a promoted behavior, can be a highly effective persuasion method (see Figure 4.5). Fear appeals tend to be the most common found in the field of emergency management public preparedness campaigns. Planners seeking to use fear appeals must be very careful, however, as these messages can and often do backfire. The following section discusses the use and viability of fear appeals in the promotion of emergency preparedness.

The Extended Parallel Process Model

A brief survey of the most common barriers to public engagement in emergency preparedness suggests that information alone is not enough to motivate the public to take action. At the end of 2006, the Council of Excellence in Government released the results of a preparedness survey of 15,000 Americans, including the most common reasons why 32% of respondents were doing *absolutely nothing* to prepare for an emergency. Among this group, the following was reported (respondents were permitted to make multiple selections):

- 45% had not thought about the issue.
- 34% did not feel that they and their families were particularly susceptible to an emergency.
- 25% believed that there was nothing that they could do to prevent an emergency if it was going to happen.
- 21% stated that they did not know what to do.
- 18% said that it took too much time.
- 16% reported that preparedness cost too much money.

In addition to the most common explanations, the report noted general themes of apathy, the assumption that emergency preparedness is a government problem (i.e., for FEMA, Red Cross, local agencies) rather than an individual problem, a lack of faith in communication and plan implementation, as well as skepticism about how effective the plan would be even if it were properly implemented (Emergency Preparedness Institute, 2007).

In essence, these results strongly support the Emergency Preparedness Institute's aforementioned suggestion that the current approach to promoting emergency preparedness is ineffective. This is not to say that the promotion of emergency preparedness is an impossible feat, but rather that the style and method of promotion needs to be reassessed. As noted by ORC Macro (2005), "While providing preparedness information appears to be important, it is not sufficient to change the behavior of most Americans. A successful campaign must also convince people that the action is important and easy to do" (p. 1). In particular, the report suggests, "The perception of an imminent threat is probably the greatest factor in motivating people to take action" (p. 1). In the field of communication science, these types of threatening messages are known as "fear appeals."

While past research indicates that messages using fear appeals have sometimes produced robust effects on attitudes and behaviors (e.g., Buck, 1984), findings have been inconsistent, and sometimes fear appeals have been shown to have little to no impact, or even generated responses opposite of the message's intent (e.g., Kohn, Goodstadt, Cook, Sheppard, and Chan, 1982). One of the most prevalent communication theories on fear appeals, Witte's (1994) Extended Parallel Processing Model (EPPM), has been highly successful in explaining such effects and providing a guideline for the construction of effective fear appeal messages.

In its most basic form, the EPPM posits that when people are exposed to threatening messages, they may respond in three different ways.

1. Those who do not feel threatened by the message are not likely to respond to it at all.
2. Those who feel threatened but do not feel that they are capable of taking action to reduce the threat will take inner-focused action to reduce their own perceptions of the risk, rather than actually taking action toward a solution. This type of reaction is generally associated with attempts to avoid thinking about the issue or an increase in the behavior that puts one at risk as a form of backlash against the threat. For example, a heavy smoker may attempt to avoid anti-smoking messages so that he or she can continue to smoke with the same regularity without having to think about it. Even worse, he or she may become defensive when confronted with such messages and respond by smoking more and claiming that everything causes cancer anyway.
3. Finally, those who feel threatened but feel that they have the capability to take the proper action to reduce or eliminate the threat will take outer-focused action toward a solution.

According to the theory, four major elements must be considered in the design of effective fear appeal messages. The first two factors have to do with the threat itself:

1. **Susceptibility**: It is essential that messages designed to scare people into taking action to protect themselves make people feel as though they are personally vulnerable to the threat being

discussed. Regarding emergency preparedness, people's perceptions that they and their loved ones were not particularly vulnerable to a threat at the moment was one of the most common reasons provided by the American public as to why they had not taken the recommended steps to prepare for an emergency.

2. **Severity**: Even if an individual is convinced of his or her own susceptibility to the threat, he or she is not likely to bother engaging the prevention behaviors being advocated if he or she does not consider the consequences of the threat as very harsh. Research indicates that emergency preparedness is considered costly in terms of time and money. In order to motivate people to take action, the benefit of avoiding the threat must greatly outweigh the cost of preparing to avoid it. It is important to note that a "severe" threat may be defined differently by different people. For this reason, it is crucial that the message designer become familiar with the target audience and emphasize the loss of something the audience values highly.

If these two elements are present in emergency preparedness messages, they are likely to be effective in scaring the target audience. Successful fear appeal messages do not stop there, however. Messages that scare the audience and provide no instruction on how to offset that fear are likely to produce the "Scarlett O'Hara" effect: "I can't think about that. If I do, I'll go crazy. I'll think about it tomorrow" (*Gone with the Wind,* Margaret Mitchell). The America Safe Foundation used the term "mental distancing" to explain that people often avoid thinking about issues that cause them stress and anxiety. This phenomenon is illustrated by the fact that the most common excuse provided for not taking steps to prepare for an emergency was that people simply "had not thought about it." The effect can be avoided by providing the audience with the confidence that there are things that they are capable of doing that will avert the threat. The final two components of the EPPM deal directly with this task.

3. **Self-Efficacy**: Solutions provided to offset a threat are only effective if people feel that they have the resources and the ability to perform the actions necessary for the solution. The claims that people do not know what they should do to prepare for an emergency and that preparing for an emergency is too costly are directly related to low self-efficacy to properly prepare for an emergency. Messages that threaten the audience must provide very clear-cut solutions that emphasize the simplicity of

the recommended actions. One way to increase self-efficacy is to show members of the target audience that people who are similar to them or with whom they identify are engaging in the recommended behaviors.

4. **Response Efficacy**: The final component of an effective fear appeal message is a convincing argument that the solution posed will be effective in offsetting the threat. Research indicates that among people who had not taken action to prepare for an emergency, there existed widespread belief that the action promoted to prepare for the emergency would probably not be effective in protecting them. This suggests that when it comes to emergency preparedness, people are not likely to take a "better safe than sorry" approach. Therefore, emergency preparedness messages must be extremely clear about how taking the steps that are advocated by the message will offset the threat.

Appropriate manipulation of these four variables has been highly effective in achieving a number of goals involving taking protective action against different threats such as AIDS (Witte, 1992), rape (Morrison, 2005), skin cancer (Parrot, Glassman, and Burgoon, 1989), and firearm hazards (Roberto, Meyer, Johnson, and Atkin, 2000).

CREATING TARGETED MATERIALS

There is no single message that will be 100% effective in promoting change among the target audience members who would benefit from such change. One segment of the population might consider a particular hazard or condition to be threatening, while another segment of the same population finds the same hazard or condition irrelevant or minor. It is for this reason that it is important to carefully identify and select the particular segment of the target audience that will be targeted with each message. The formative research conducted during earlier campaign planning stages, such as the holding of focus groups, provides valuable insight into what types of things the segment finds threatening and what kinds of barriers stand in the way of behavioral change geared toward threat reduction.

Generally, campaign message targeting decisions are based on demographic information such as age, income, gender, or education level. It is important to remember that such divisions rarely guarantee uniform results among the targeted population. Two theories often used in communication and social psychology can help to provide insight into other factors that may separate people's responses even within the same targeted demographic. These include the Transtheoretical Model of Intentional Behavior Change and the Diffusion of Innovations (DOI) Theory.

- **The Transtheoretical Model of Intentional Behavior Change** — The transtheoretical model of intentional behavior change (Procheska and DiClemente, 1983) suggests that individuals make decisions about behavior modifications through a series of behavior-change transitions called the "stages of change." These stages consist of:
  - **Precontemplation:** People in the precontemplation phase have no intention of changing their behavior regarding the issue being promoted within the next six months.
  - **Contemplation:** Those in the contemplation phase still have no solid plans to make any immediate changes, but are starting to think about making behavioral changes involving the issue being promoted sometime within the next six months.
  - **Preparation:** Those who have moved from the contemplation phase to the preparation phase have not yet changed their behavior regarding the issue being promoted, but they plan to within the next 30 days.
  - **Action:** People in the action phase have taken behavioral action regarding the issue being promoted, but this change in behavior has occurred within the last six months. As a result, the behavior is not permanent; enactment still requires effort. Many emergency preparedness messages need to promote a single-time behavior, such as putting together an emergency preparedness kit or temporarily evacuating the home. Obviously, then, if the behavior to be promoted only requires that people take a single action, the end goal is bringing people into the action phase.
  - **Maintenance:** Sometimes the action being promoted requires that the public consistently perform the action over time, such as updating the family preparedness plan as children get older and daily schedules change. In this case, the behavior being promoted needs to become a habit. The end goal of these types of messages is to bring people beyond the action phase

into the maintenance phase in which people have consistently performed the action and it is now part of their routine.

Practitioners can benefit from using these stages as a guideline for making decisions about targeting their messages and for setting realistic goals for the messages to achieve. For example, if a survey of the population of interest indicates that the vast majority of people know nothing about the behavior being advocated, they should probably be classified as being in the precontemplation phase. Depending upon people's pre-existing attitudes toward the behavior and the number and size of the barriers that may prevent people from taking action, it may be realistic to believe that a set of messages could inspire people to move from the precontemplation phase to the contemplation phase, the preparation phase, or possibly even the action phase. In most cases, however, it is unreasonable to expect that a single set of messages promoting emergency preparedness is likely to move people from the precontemplation to the maintenance phase. Therefore, if pre-test measures suggest that most people in the target audience are in one of the earlier stages of change, it may be wise to plan on the circulation of multiple messages over time with smaller goals of moving people from the precontemplation stage to the contemplation stage or from the contemplation stage to the preparation stage.

This insight also has implications for message design. Obviously, messages with the goal of moving people from the stage of precontemplation to contemplation are going to be different from those attempting to push people from preparation to action. In the precontemplation and contemplation stages, it is likely to be important to raise awareness about the issue at hand and convince people that the issue is worthy of their time and consideration. Once the importance of the issue has been established and people have moved from the precontemplation to the contemplation and preparation stages, messages promoting action should emphasize the degree to which the benefits of the action outweigh the barriers associated with it. These types of messages should also promote self- and response-efficacy — the beliefs among target audience members that they have the ability to perform the action being promoted and that this action will successfully offset the dangers associated with possible emergencies.

- **Diffusion of Innovations Theory** — Helpful insight about message targeting decisions can also be drawn from the work of communication scholar Everett Rogers (1983). Rogers developed a theory about how different norms, ideals, and practices come to be adopted or rejected by society called Diffusion of Innovations. While the theory itself is complex and beyond the scope of this book, one specific element discussed within the theory can aid in determining what types of people the messages should be targeted toward and in setting realistic expectations about the adoption of emergency preparedness behaviors over time. This element deals with the social system through which new ideas (i.e., innovations) are either adopted or rejected. The theory suggests that successful new technologies or practices come to be known by different segments of the population at different times. These segments are as follows:

- **Innovators**: Innovators are generally the first people to ascribe to new technologies or behaviors. This segment of the population is composed of a small number of nonconformists who generally do not have much influence over the actions of the rest of the population. Innovators are said to make up only 2.5% of the population.
- **Early Adopters**: The people who are said to have the largest impact on whether an innovation catches on among the general population are generally included within this segment of the population. These people are often referred to as *opinion leaders* because of the role they play in shaping public opinion. Opinion leaders are often more media savvy and well-connected members of the population. These people are most likely to be influenced by mass media communication campaigns. Most others are likely to be influenced secondhand through interpersonal channels of communication. Therefore, when designing a message, it is important to identify characteristics of opinion leaders within the target population to create messages that may appeal to them specifically. Early adopters compose 13.5% of the population.
- **Early Majority**: Innovators, early adopters, and laggards (the last category) make up a very small portion of the population. The remaining majority of the population is divided according to the degree to which they are resistant to change. The early majority are those members of the majority who are more open to adopting innovations. The adoption of the new technology or behavior by the early majority sets into motion more and more people embracing the behavior until it reaches *critical mass*, the point at which most people have adopted the

innovation and others begin to feel a need to catch up. The early majority is 34% of the population.

- **Late Majority**: Members of the late majority are highly resistant to change. It generally takes a significant event such as a pressing need or a critical mass that finally leads this segment to adopt a new behavior. The late majority accounts for 34% of the population as well.
- **Laggards**: Laggards are the least likely segment of the population to adopt a new behavior. Many laggards never change. Those who actually adopt a new behavior do so generally because their current behavior is unavailable. Laggards make up 16% of the population.

Knowing these typical patterns of adaptation can be informative for campaign designers in that they acknowledge the reality that some people are more apt to try new things and some are unlikely ever to be swayed regardless of the messages provided to them. The key to successfully promoting new messages is to identify the opinion leaders of the target population and tailor messages to appeal to them.

SOCIAL NORMS

Once a critical mass has been reached within which most people either are engaging in a promoted behavior themselves or are at least approving of others engaging in the behavior, another appeal that is often effective is social norms. In the fields of communication and social psychology research, the theory of reasoned action (TRA; Ajzen and Fishbein, 1980), focus theory (Cialdini, Reno, and Kallgren, 1990; Reno, Cialdini, and Kallgren, 1993), the theory of normative social behavior (TNSB; Rimal and Real, 2005; Lapinski and Rimal, 2005), and norm activation theory (Schwartz, 1973) address the influence of social norms on behaviors. Most of the literature discusses one or more of three different types of potentially influential social norms:

1. **Subjective Norms**: Subjective norms refer to a person's perception of what others who are important to the person expect him or her to do. A message used in an attempt to persuade with a subjective norm should seek to remind or reveal to the target audience that the people closest to them, such as family and friends,

expect them to engage in the behavior being promoted. For example, an emergency preparedness campaign might include messages reminding the target audience of their children's expectations for them to have a plan to keep them safe in the event of an emergency.

2. **Injunctive Norms**: Rather than just considering what important others expect, injunctive norms involve the types of behaviors of which important others approve and disapprove. It is generally expected that behaviors that are in line with the injunctive norm will be socially rewarded and behaviors that go against the injunctive norm will be socially punished. A message used in an attempt to persuade with an injunctive norm should seek to remind or reveal to the target audience that the people who are most important to them approve of or disapprove of the behavior in discussion. For example, an emergency preparedness campaign might include messages revealing to the target audience the percentage of their peers who think that they ought to talk to their children about the plan of action in the event of an emergency.
3. **Descriptive Norms**: Descriptive norms refer to the prevalence of a behavior within a population. It has been shown that people tend to perceive that certain unhealthy behaviors are far more common in a population than they actually are. For example, when college students are asked to estimate the percentage of students at their university that have five or more drinks in one sitting, they often estimate that the vast majority of students engage in this behavior. In reality, however, less than half of students actually do partake in this type of excessive drinking. The widespread incorrect perception of how common it is for students to drink five or more drinks in one sitting is thought to encourage other students to increase their own drinking in order to be "normal." The use of messages to clarify that drinking in excess is far less common than they think has been shown to reduce the number of students who engage in this type of behavior. In terms of emergency preparedness, people may feel that their failure to take the proper actions to prepare themselves and their families for a disaster is excusable because most other people have not prepared themselves or their families either. If people were informed that most other people have, in fact, taken the proper emergency preparedness actions, they are likely to feel pressure to take these behaviors themselves in order to be normal.

DESIGN MATERIALS

Campaign designers should use the insight provided by theory as well as results of formative research that was conducted in the early stages of campaign development to create targeted messages to be used within the campaign. The fact that a message appears to contain all of the elements of the theory on which it is based and was derived from the results of formative research with members of the target audience, however, does not necessarily mean that it will be effective in bringing about the desired change among the target audience. For this reason, it is crucial for designers to brainstorm and generate a number of potential campaign messages to be pre-tested in a focus group setting with more members of the target audience.

ACTIVITIES AND EVENTS PLANNING

Many of the communication channels described in this chapter involve participatory activities or events to achieve message transmission. Events and activities are beneficial in that they allow communicators the opportunity to extend the reach of their campaign across many audience members at once. However, this adds a certain level of complexity because of the planning that must be conducted in advance of the activity or event to ensure that it achieves the desired outcome. Event planning is a distinct skill set that should be considered when forming the planning team. As was true with the selection of communicators, planners must consider the preferences and needs of the audience in the planning for risk communication events and activities.

The following list provides a brief overview of the issues that must be considered when planning events or activities:

- Should the event or activity be stand-alone (e.g., a demonstration at a local elementary school), co-hosted by other organizations with similar interests (e.g., a public safety fair), or should the organization piggyback a larger event that is being planned and conducted by another organization (e.g., a county fair)?
- When should the event be held, at what time of day, and for what duration?
- What budget exists or should be allocated for the event?
- How many participants or attendees are expected to come, what is their demographic makeup, and what are their expectations and reasons for attending?

- How will the event be advertised (press releases, or print, television, and radio ads), or will invitations be used?
- How will you ensure publicity for the event? Will the press be invited and what information will you provide them (e.g., press kits)?
- What type of venue is required to host the event (including size, accessibility, services, equipment, etc.), and what options exist in the community?
- How will the event be managed?
- What will the staffing needs be (workers and volunteers), what skills will be required, how will schedules be managed, and what will be the standard operating procedure for staff and volunteers interacting with the public?
- What speakers will be invited and what will they talk about?
- What activities will take place and how will they be conducted?
- What risk reduction educational materials will be created and distributed and how many are needed?
- Will food and refreshments be served?
- What supplies (e.g., giveaways, paper and pens) and equipment (e.g., audio/visual, communications) are needed to conduct the event or activity?
- What liability coverage is required, and what exists or is being done to ensure the safety of all in attendance (emergency exits, law enforcement, and emergency medical care)?
- Is access to all participants with special needs ensured (see http://www.health.state.ny.us/nysdoh/promo/events.htm for more information)?
- How can participants be tracked after the event?
- What are the expected outcomes of the event or activity, and what methods for evaluation exist?
- What follow-up will be conducted (see http://www.nationalwaterweek.org.au/guide-to-running-events.php)?

ESTABLISHING A PROJECT TIMETABLE

One of the last project management tools created before campaign implementation begins is the project timetable, which details both implementation and evaluation of the campaign. Communication campaigns involve dozens, if not hundreds, of individual tasks that must be carried out. Timelines allow for the priority ranking of tasks, and help to illustrate tasks across the life of the project (making other management problems such as staffing and funding easier to assign). The timeline is most effective if it includes every task that will be carried out, no matter how small. By including all tasks, project managers help to ensure that they are able to carry out all tasks at the time they are needed, to assign workers/volunteers and resources to only one task at a time, and to make sure that the project is kept on its intended schedule. Timetables should be created in a way that allows for mid-campaign changes, and additions or deletions of individual tasks that become irrelevant as the project is implemented. In addition, because timelines are fluid in nature, they should be reviewed periodically and adjusted to account for any changes that have occurred since the last review. Timelines are included in the various project management software packages described in Chapter 3.

CREATING THE COMPREHENSIVE COMMUNICATION PLAN

The culmination of project planning is marked by the creation of a comprehensive communication plan. The communication plan provides those working on the campaign with an accurate blueprint of how the intended communication will play out, and provides for all partners a tangible and established record of what will be taking place and what is expected to occur as a result. The communication plan also allows communicators to explain easily their campaign to those outside the campaign. This document presents and explains the project's budget, and justifies spending where required. Finally, it illustrates the lifecycle of the campaign. There are several components normally included in a comprehensive communication plan, as follows:

- The campaign strategy (described at the beginning of this chapter)
- Partnering plans
- Message and materials development and testing plans
- Implementation plans (including standard operating procedures and plans for distribution, promotion, and process evaluation [described in Chapter 5], for example)
- Outcome evaluation plans
- Task lists and timelines

PILOT TESTING AND ADJUSTING CAMPAIGN MATERIALS

It is highly difficult to predict which campaign materials will be most liked or disliked by target audience members. Therefore, once campaign materials have been generated, it is necessary to conduct additional focus groups to pilot test them all before making any decisions about their use in the campaign. These focus groups should be conducted in a fashion similar to those conducted before the start of the campaign development process. This time, however, rather than asking questions about group members themselves, group moderators should probe for group members' thoughts about each of the potential campaign messages. Some salient questions to ask may include:

1. What do you think about when you see/hear/read this message?
2. Which message do you like best, and why?
3. Which message do you like least, and why?
4. Which font/color scheme do you like best, and why?
5. Which font/color scheme do you like least, and why?
6. Would you be motivated by this message to engage in the prescribed behavior?

Pre-tests can be highly informative in weeding out messages that are generally despised, despite being guided by theory and derived from the formative research on target audience members. Moreover, even messages that are highly regarded by focus group members can still be improved upon. The purpose of the pilot test meeting is to identify which of the messages would be most effective among group members and what changes can be made to the message to make it even more effective. Based on the results of this exercise, campaign designers should select a single message design, make any final adjustments that may have been suggested within the group, and prepare for message distribution.

CONCLUSION

The development of a campaign strategy and a communication plan requires as much creativity as it does theory and the application of tested methodologies. At the completion of this step, the planning team will have determined how they will communicate with their audience, including the settings, channels, and methods. Their campaign strategy will be set, as dictated by the market research they conducted in preparation for the risk communication efforts to follow. Chapter 5 describes the process by which all of this preparation becomes actual communication, and the ways in which communicators measure their effectiveness.

REFERENCES

Ajzen, I. and Fishbein, M. 1980. *Understanding attitudes and predicting social behavior.* Englewood Cliff, NJ: Prentice Hall.

Buck, R. 1984. *The Communication of Emotion.* New York: Guilford Press.

Cialdini, C. A., Reno, R. R., and Kallgren, C. A. 1990. A focus theory of normative conduct: Recycling the concept of norms to reduce littering in public places. *Journal of Personality and Social Psychology, 58,* 1015–1026.

Cockerham, William, G. Lueschen, G. Kunz, and J. Spaeth. 1986. Social stratification and self-management of health. *Journal of Health and Social Behavior*, 27, 1–14.

Covello, Vincent, and Peter M. Sandman. 2001. Risk Communication: Evolution and revolution. In *Solutions to an Environment in Peril.* Anthony Wolbarst, Ed. Baltimore, MD: Johns Hopkins University Press, pp. 164–178.

Emergency Preparedness Institute. 2007. *Preparedness Needs a New Message.* Mt. Pleasant, IL: Emergency Preparedness Institute.

Fuller, M. 1991. *Forest Fires.* New York: John Wiley & Sons.

Government of Canada. 2009. Active2010. Government of Canada Website. http://www.active2010.ca/

Kohn, P. M., M. S. Goodstadt, G. M. Cook, M. Sheppard, and G. Chan. 1982. Ineffectiveness of threat appeals about drinking and driving. *Accident Analysis and Prevention,* 14, 457–464.

Lapinski, M.K and Rimal, R.N. 2005. An explication of social norms. *Communication Theory, 15,* 127–147.

Morgan, M. G., B. Fischoff, A. Bostrom, and C. J. Atman. 2002 . *Risk Communication: A Mental Models Approach* . Cambridge: Cambridge University Press.

Morrison, K. 2005. Motivating women and men to take protective action against rape: examining direct and indirect persuasive fear appeals. *Health Communication, 18*(3), 237–256.

National Research Council (NRC). 1989. *Improving Risk Communication.* Washington, D.C.: National Research Council, National Academies Press.

NIH. 2009. Healthy People 2010. NIH Website. http://www.healthypeople.gov/

ORC Macro. 2005. Citizen preparedness review. *Citizen Corps.* Issue 1. Washington, D.C.

Parrott, R., Glassman, M., & Burgoon, M. 1989, November. *Arizona's public campaign to stop overexposure to the sun.* A paper presented in the Applied Communication Division of the Speech Communication Association's annual meeting in San Francisco, California..

Procheska, J. O., and C. C. DiClemente. 1983. Stages and processes of self-change in smoking: Toward an integrative model of change. *Journal of Consulting and Critical Psychology,* 5, 390–395.

Reno, R. R., Cialdini, R. B., and Kallgren, C. A. 1993. The transsituational influence of social norms. *Journal of Personality and Social Psychology, 64,* 104–112.

Rimal, R. N., and Real, K. June, 2005. How behaviors are influenced by perceived norms: A test of the theory of normative social behavior. *Communication Research, 32*(3), 389–414.

Roberto, A. J., Meyer, G., Johnson, A. J., and Atkin, C. K. 2000. Using the extended parallel process model to prevent firearm injury and death: Field experiment results of a video-based intervention. *Journal of Communication,* 157–175.

Rogers, E.M. 1995. *Diffusion of innovations (4th edition).* The Free Press. New York.

Schwartz, S. H. 1973. Normative explanations of helping behavior: A critique, proposal, and empirical test. *Journal of Experimental Social Psychology, 9,* 349–364.

Swanson, Jordan. 2000. Unnatural disasters. *Harvard International Review,* 22(1), 32–35.

UNESCO Press. 2002. *Literacy as Freedom.* UNESCO Institute for Statistics. http://unesdoc.unesco.org/images/0013/001318/131823e.pdf

Witte, K. 1994. Fear control and danger control: A test of the extended parallel process model (EPPM). *Communication Monographs,* 61, 113–134.

World Bank. 2002. *Natural Hazard Risk Management in the Caribbean.* Latin American and the Caribbean Region: Author.

5

Campaign Implementation and Evaluation

INTRODUCTION

Upon completing all of the background research, design, and planning, it is finally time for the project team to begin implementing its campaign. In Chapter 3, the importance of setting measurable goals and objectives was stressed. Meeting those objectives is the entire purpose of the campaign, and by assessing them directly, the campaign team is able to determine whether those goals have been met or whether the campaign was a success. One can never use arbitrary or anecdotal evidence to assume the triumph of a team's efforts; only properly planned and conducted evaluation can measure the impact a campaign has had on an audience. This chapter examines the launch, management, and evaluation of an emergency preparedness public education campaign.

CAMPAIGN LAUNCH

A program's launch sets the stage for the rest of the campaign to follow. It is therefore crucial that planning teams have previously taken the time to consider as many of the roadblocks the campaign may face and have already formulated plans of action to avoid what is preventable (see Sidebar 5.1). While in-process changes are possible, the campaign team should always set out to achieve its goals with the campaign it launches.

FIGURE 5.1 New Orleans, Louisiana, May 21, 2006 — Wali Armstead, FEMA Public Affairs Officer, hands a preparedness kit to raffle winner Mary Rogers at the "FEMA Family Preparedness Day" in New Orleans. FEMA brought together many organizations to present emergency preparedness information to allow parents and children to gather emergency supplies and decide where they will find safe shelter for themselves and pets during dangerous storms. Many of the items that go into an emergency supply kit were provided free at the event during raffles, activities, and games (Marvin Nauman/FEMA photo).

Many programs begin with a *kickoff event*. Like the kickoff meeting described in Chapter 4, a kickoff event is designed to draw attention — in this case to the launch of the campaign. A campaign cannot successfully persuade its audience if the members of that audience are not exposed to its messages. Therefore, the kickoff event is used to garner media coverage of the campaign, which in turn creates exposure. Campaign planners must keep in mind that as significant as they believe the message of their campaign to be, it will be but one of thousands of critical issues competing for news coverage every day. Therefore, it is vital that they create something exciting and new that grabs people's attention (see Figure 5.1). Whatever the event, it should always be closely tied to the campaign's issue and allow ample opportunity to expose people to the campaign's message.

A well-run kickoff event produces what is referred to as *brand recognition*. Brand recognition helps audience members to pay attention to and later recall campaign messages when they encounter communicators and their methods. A kickoff event can also generate increased community involvement in the espoused cause, which in turn may lead to an increase in interpersonal conversations about the issue being promoted. All of these generally lead to a positive change in behavior.

THE MEDIA

In order to inform the press about a kickoff or other event, or about the campaign in general, campaign designers create media kits that are distributed to various local and national media outlets. A media kit is a collection of information about the campaign that generally includes things such as:

- A press release about what is being promoted
- Research and statistical information about the issue as it pertains to the target audience of the media outlet, complete with charts, graphs, and other visuals that can be used in a news story
- Background information about the problem and the proposed solutions
- Credibility-enhancing background information about those associated with the campaign

Media kits are relatively simple to make, but it can quickly become expensive if they are developed for every outlet. Therefore, it is highly cost-effective to research local media outlets before considering which to approach. It would be wasteful, for instance, to send a media kit to a newspaper that clearly never covers any issues that are similar to the one promoted by your campaign. Among the outlets that are decidedly viable options to receive media kits, it is also important to consider that it is not a guarantee that your media kit will be read. If possible, it is a good idea for a spokesperson from the campaign to hand-deliver all media kits directly to people working at the media outlet.

Sidebar 5.1: Launch Checklist

It is always important to consider all possibilities and to be prepared for things that may not work according to plan. This is particularly important when the campaign launch includes a kickoff event that is going to be covered by media outlets. The NIH's "Planner's Guide to Making Health Communication Programs Work" provides a checklist, in the form of questions, that can be used before launching a campaign to ensure that everything about the campaign's launch has been considered. Items on this checklist include:

- Are our partners prepared for the launch?
- Have we invited all the necessary partners and stakeholders who have been involved in program development?
- Have we prepared (or trained, if necessary) our staff and spokespeople?
- Are program-related services (e.g., a hotline, inspection service) in place?
- Do we have a list of media outlets we need to contact?
- Are all of our promotional materials ready?
- Do we have enough materials to start the program (e.g., PSAs and media kits) and respond to inquiries (e.g., leaflets for the public)?
- Are reordering mechanisms in place?
- Do we have mechanisms in place to track and identify potential problems?
- Are emergency managers and other emergency service providers in the community aware of our program and prepared to respond if their constituents ask about it?

EVALUATION

In their chapter, "Communication Campaign Effectiveness: Critical Distinctions" in Rice and Atkin's (2001) widely cited book, *Public Communication Campaigns,* Salmon and Murray-Johnson draw a distinction between campaign *effectiveness* and campaign *effects.* They state that an *effective* campaign is one in which the goals and objectives set during the preliminary stages of campaign design are met. Campaign *effects,* on the other hand, are any outcomes produced by the campaign. Not all campaign effects contribute to campaign effectiveness. In fact, in some cases campaign effects can be contrary to the campaign's purpose. For example, in the public health sector a systematic review of published studies evaluating substance abuse prevention programs saw evidence that, while some programs resulted in reduced levels of drug and alcohol consumption, and some had no impact on substance use at all, others appeared to *increase* drug and alcohol use among target audience members (Werch and Owen, 2002). There are various reasons that explain why this might occur, but it is suggested a primary cause is criticism of the campaign and its efforts by people influential to the target audience members. In such cases, a campaign may be highly effective in generating message exposure, but is considered *ineffective* because it produces an *effect* that is opposite its intended objective. Proper evaluation needs to detect contradictory effects and stop them from spreading.

In addition to preventing further harm, evaluation improves the odds that a program will be successful. Using evaluation methodologies, campaign designers are forced to set measurable objectives that the campaign will attempt to achieve. Once established, these decisions may be used to guide all factors relevant to the design process. Materials, events, and messages are each created and designed according to the goal of meeting these predefined objectives.

OBJECTIVITY IN EVALUATION

After spending large amounts of time and energy creating campaign materials, it can be impossible to provide a truly objective viewpoint on the project using one's own personal assessment of success. Even the most unbiased individual is unlikely to succeed in observing a strictly representative sample of the target audience in order to ensure that observed effects are uniform throughout the target population. Therefore, the final campaign assessment should ideally be based on trying to achieve preset effects that are objectively measurable.

Based upon the three primary public disaster preparedness education goals, namely, (1) raising awareness of the hazard/risk(s); (2) guiding public behavior, including pre-disaster risk reduction behavior, pre-disaster preparedness behavior, post-disaster response behavior, and post-disaster recovery behavior; and (3) warning the public, some examples of measurable effects may be

1. Increasing the total number of hits on a Website
2. Increasing scores on a knowledge test distributed among the target audience

3. Increasing sales of safety or preparedness equipment being promoted
4. A count of the number of people engaging in the promoted behavior obtained through some type of random selection such as random digit telephone calling

Objectives based on these goals could include, for example:

1. Increase by 20% the number of hits on the campaign's Website within two months.
2. Increase the average score on a knowledge test administered to high school students by five points.
3. Increase the total number of people using tornado-resistant safe rooms in the target area by 10%.

THE JUSTIFICATION FOR EVALUATION

In the end, campaign success is determined not by the number of people who merely were exposed to the message, nor by the degree to which people "like" or "enjoy" the campaign materials or messages, but rather by the degree to which it meets the objectives set at the beginning of the campaign design process.

Campaign assessment is not a simple task. Most of the statistical techniques required to achieve a useful evaluation of campaign effect sizes are well beyond the scope of this book. For this reason, it is highly recommended that campaign evaluation be left to professionals with statistical expertise and training relevant to public education campaign design. Hiring a professional with such knowledge can cut heavily into the overall campaign budget, which is a primary reason why project planners often neglect evaluation. Campaign planning teams may reason that it is senseless to spend money on evaluation that could have otherwise gone toward extending the campaign's reach and increasing exposure among the target audience. The response to such accusations is that campaign evaluation is the *only way* to ensure that the campaign is effective in meeting its objectives. Perhaps more importantly, without evaluation there would be no way to ensure that the campaign is not producing any effects that run *counter* to its objectives. Finally, if campaign evaluation indicates that the messages have been highly effective, the results can be used to garner more funding to continue with the campaign to extend far beyond its initial reach.

Launching a campaign is generally both an exciting and a stressful event. It can be fun to see a vision, created through much research and hard work, transformed into reality. Still, it is crucial to recognize that no matter how well prepared, organized, and straightforward the implementation plan may be, and no matter how responsible and well-trained campaign staff is, it is almost certain that not everything will go according to plan when it comes time for implementation. For this reason, it is vital to monitor the ongoing campaign through process evaluation (see the next section).

PROCESS EVALUATION

Process evaluation is a procedure through which researchers assess the degree to which campaign implementation is being conducted as planned. Process evaluation is conducted at specified points throughout the campaign. For instance, it would be impossible for brochures that were designed to provide the public with information about local natural hazards to do so if they were never actually distributed to the target audience. Likewise, promotional events would be unlikely to have the intended impact if the target audience was not made aware of them through proper advertising methods. Throughout the course of a campaign, countless details can be forgotten, misunderstood, or misinterpreted by campaign staff or audiences. A proper process evaluation helps to measure these effectiveness issues by answering the following three questions:

1. **Is the target audience being exposed to the message?** As was previously noted, a campaign cannot effectively meet its objectives if the target audience does not see or hear its messages. The process evaluation effort should therefore include a measurement of campaign exposure among members of the intended audience.
2. **Is the campaign producing any unanticipated adverse effects?** No matter how much formative research or theory is used to guide campaign design, there is no way to predict how an audience will interpret a message. Even the most well-intentioned message can be misinterpreted or disliked by the audience for reasons not understood by the communicators. Generally, focus groups that are used in formative research fail to form a representative sample of target population members, which in turn greatly

increases the likelihood that opinions and tastes of the general target populations will be misjudged. Process evaluation assesses overall reactions among the intended audience. For instance, process evaluation results may suggest that campaign messages are offensive, misunderstood, disliked, or simply ignored, giving communicators a chance to adjust mid-stream and avoid outright campaign failure.

3. **Is the campaign being conducted exactly as planned?** Unforeseen roadblocks undoubtedly appear during the course of campaign implementation. Process evaluation helps to correct that which can be fixed and documents that which cannot. This information provides communicators with an opportunity to assess their outcomes in response to what *is being done,* as opposed to what *was planned to be done,* giving them the chance to make more knowledgeable process adjustments.

Communicators should keep in mind the knowledge about how a campaign is conducted when making assertions about the campaign's effectiveness at its conclusion. It is crucial to remember that all of the effects that are detected in campaign evaluation should be attributed to the actual campaign that was carried out rather than the campaign that was planned. Information about these effects comes from the campaign's *summative evaluation.* Although summative evaluation is not actually conducted until the conclusion of the campaign, it plays a large role in decisions made throughout the entire process. The following sections detail various campaign design decisions to be made and the implications they have for final campaign evaluation.

EXPERIMENTAL AND QUASI-EXPERIMENTAL DESIGN

In a perfect world, all campaign summative evaluations would stem from perfectly executed *true experimental designs.* True experimental designs are the only way to rule out all other explanations for the effects produced by the campaign effort. The following section defines experimental design and distinguishes it from simpler designs called quasi-experiments. The most common types of experiments are listed. In addition, suggestions are made about the practicality of each as it pertains to your own campaign.

True Experimental Designs

A true experimental design is the only way to determine conclusively that the effects produced are due to the campaign and the campaign alone. All other designs open themselves up to a variety of criticisms and alternate explanations for the effects produced. True experimental designs require that three conditions be met:

1. **Treatment Group:** The treatment group is the segment of the population that receives the campaign that has been designed to produce changes in knowledge, attitudes, and behavior.
2. **Control Group:** The control group is the segment of the population that is the same as the treatment group in all ways except for the fact that it does not receive the campaign messages that were designed to produce changes in knowledge, attitudes, and behavior. The control group provides a marker of what would be happening throughout the entire population had it not been for the campaign. Keeping all things (i.e., demographics, living conditions, location, etc.) except for the campaign constant allows for the attribution of all differences between the treatment and the control group at the conclusion of the campaign to the campaign itself.
3. **Random Assignment to Conditions:** Random assignment to conditions means that every person to be assessed in the final evaluation has an equal chance of being assigned to either the treatment or the control group. Statistically speaking, use of random assignment provides the greatest probability that all things, including demographics, living conditions, location, and others, are evenly distributed across the treatment and control conditions. Random assignment to conditions is the only method that allows researchers to assume that all things besides the campaign itself were held constant between the treatment and control groups and therefore any differences between the treatment and control groups can be attributed to the campaign.

When using random assignment, it is important to remember that "random" does not mean "chaotic." A legitimately random assignment technique gives everyone involved an equal probability of being assigned to either the control group or the treatment group. Evaluators can do this using traditional or computerized randomization models that range from picking slips of paper out of a hat to highly complex number-generating software.

Random assignment is what differentiates true experimental designs from quasi-experiments (a term introduced by Campbell and Stanley [1963]), which are discussed later in this chapter. There are three types of true experimental designs, the benefits and drawbacks of which are described in the following. Communicators should be aware, however, that these methods and examples are presented in a simplistic manner, and that the results they generate do not offer as clear effects as could be claimed by experimenters who apply rigorous experimental design and statistical data testing. Moreover, while it is not always feasible, trained professionals should evaluate campaigns for this reason. The three types of true experimental designs are

1. **Post-Test Only Control Group Design**: This method is the simplest of true experimental designs. This design assesses all participants in both the treatment and the control groups in terms of the independent variable at a single point in time after the campaign has completed. Randomly assigning all participants to either the control group or the treatment group allows researchers to assume that their scores on the outcome variable before campaign implementation are roughly even. Because the control group is not exposed to the campaign, it is expected that its outcome scores will remain constant throughout the campaign. Those in the treatment group, however, should change outcome scores as a factor of campaign exposure such that they meet the objective set in the campaign's early design. The post-test scores of those in the control condition can be used as an indicator of what the scores would be for the treatment condition before the campaign, and any difference between the treatment and control group in terms of scores on the outcome variable can be attributed to the campaign. Therefore, for example, if campaign planners set out to measure their success in increasing emergency preparedness knowledge among high school students by 20%, they might use the following steps:
   a. Recruit several high schools that agree to participate in the study.
   b. Use random assignment to create control and treatment groups. Because the campaign would likely be conducted within the schools themselves, preventing those in the control group from being exposed to the campaign would be difficult. Therefore, changing the *unit of analysis* from the individual (student)-level

to the group (school)-level would yield results that are more accurate. This change would mean that each *school*, rather than each *student*, would be assigned a unique number, and all students in the schools whose numbers were randomly selected would be considered part of the treatment group.

c. Administer the campaign within all schools in the treatment condition and not in the schools in the control condition.
d. After the conclusion of the campaign, assess all students in both conditions in terms of the outcome variable. In this case, the objective is to increase knowledge about emergency preparedness among high school students. Knowledge may be assessed through a test that may be administered to all students in both the treatment and the control conditions. The resulting scores from this test serve as the post-test measure of the outcome variable of importance. It is important to remember that if the unit of analysis is at the school level, the post-test score of the treatment group will be calculated by averaging together *the average test score of each school* in the treatment condition, and the post-test score of the control group will be calculated by taking the average of the *average scores of each school* in the control condition. If the unit of analysis had been at the *individual* level, then the post-test scores could be obtained by taking the average of individual scores of people in each condition. However, because it was schools that were randomly assigned to conditions rather than individuals, the average score from each school is the lowest unit of analysis permitted.
e. Compare the average post-test scores of those in the treatment group to average post-test scores of those in the control group. As noted previously, because random assignment was used it can be assumed that before the campaign the two groups were equivalent in their emergency preparedness knowledge. Then the treatment group received the campaign and the control group received nothing. Therefore, any differences between the groups in terms of emergency preparedness knowledge can be attributed to the campaign itself.

If the average test scores of schools in the treatment condition are 20% higher than the average test scores of the schools in the control condition, then we can assume that the campaign was successful in achieving the objective to increase emergency preparedness knowledge by 20%. In this case, it would then be wise to conduct the same campaign in the schools that served as the control group during the experiment. If the average test scores of schools in the treatment condition are not 20% higher than the average test scores of those in the control condition, then it can be concluded that the campaign did not meet its objective and should therefore not be implemented elsewhere until campaign adjustments have been made.

All true experimental designs allow researchers to determine whether the campaign had an adverse affect on the outcome variable, as was the case in some of the public health campaigns discussed at the beginning of this chapter. If the average post-campaign test scores of schools in the treatment condition are actually lower than average test scores of schools in the control condition, researchers must conclude that the campaign in fact moved the population of interest further away from the campaign's goals and that they would have been better off if no campaign had been conducted at all. Obviously, under such conditions, it would be irresponsible and unethical to continue with the campaign or to implement the same campaign in any additional communities in the future.

The post-test only design is the simplest of all true experimental designs. It is more efficient and less costly than other experimental designs because it requires data collection at only one single point in time. However, equalization through randomization is only a statistical probability — not a guarantee. Therefore, there is always a possibility that even after randomization starting scores on the outcome variable differ across conditions. If this were the case, it would not have been appropriate to attribute to the campaign all differences in scores on the outcome variable after the campaign has concluded. Doing such could result in incorrect inferences about the campaign's effectiveness. For example, if after randomization, the average emergency preparedness knowledge score in the treatment group was 20% higher than the average emergency preparedness knowledge score in the control condition before the campaign, then after the conclusion of the campaign, if the difference in knowledge scores between the two groups is still only 20%, the campaign would actually have had no effect on the outcome variable. If we had assumed the two groups' scores to be equal to begin with, we would have hailed the campaign's success and recommended investing time and money into continuing with an ineffective campaign and implementing it in new areas beyond the study. These concerns often drive planners to choose more complicated designs, such as the pre-test post-test control group design discussed next.

2. **Pre-Test Post-Test Control Group Design**: As suggested by its name, a pre-test post-test control group design is an experimental design involving a randomly assigned control group and a randomly assigned experimental group, both of which are assessed in terms of the outcome variable before the campaign starts as well as at the conclusion of the campaign. If we choose to use a pre-test post-test control group design instead of a post-test only control group design for the example about increasing emergency preparedness knowledge among high school students provided previously, much of the process would remain the same:
   a. Recruit a number of high schools to agree to participate in the study.
   b. Use random assignment techniques to create control and treatment groups. It would still make sense to use schools as the unit of analysis, rather than individual students.
   c. Rather than making the assumption that random assignment equalized emergency preparedness knowledge scores across treatment and control conditions as was done with the post-test only design, the pre-test post-test control group design allows planners to ensure that scores are comparable before the start of the campaign by taking a measure of knowledge across both conditions before campaign implementation. In order to do this, one would assess all students in both conditions in terms of the outcome variable using the same test that will be used for the post-test after the campaign's completion. The resulting scores from this test serve as a pre-test measure of the outcome variable of importance. As noted, because schools have been randomly assigned to condition, the difference between scores produced by schools in the treatment condition compared to those produced by schools in the control condition should be negligible in the pre-test measurement. If this is not the case, it will be inappropriate to attribute all differences in knowledge scores between the treatment group and the control group to the campaign.

d. Once pre-test measures have been taken, planners administer the campaign in the treatment group schools and do nothing in the control group schools, just as occurred in the post-test only design.
e. At the campaign's conclusion, Step c is repeated to calculate post-test scores for both conditions. By comparing the pre-test score of the treatment condition to the post-test score of the control condition, researchers can determine whether the campaign has met its objective of increasing emergency preparedness knowledge by 20%. The pre-test post-test control group design also allows us to determine whether increases or decreases in knowledge are, in fact, due to the campaign, or whether they are due to some other external factor by comparing changes in knowledge over time in the treatment condition to changes in knowledge over time in the control condition. Even if knowledge increases in the treatment condition from pre-test to post-test, if the post-test knowledge scores in the control condition are comparable to the post-test knowledge scores in the treatment condition, then the knowledge gain cannot be attributed to the campaign. In addition, if the post-test knowledge scores in the control condition are higher than are those in the treatment condition, researchers must consider the possibility that the campaign may have had a negative impact. This result suggests that students would have been more knowledgeable about emergency preparedness if they had not been exposed to the campaign.

While pre-test post-test control group designs are better able to confirm that differences between the treatment and control groups at post-test are completely attributable to the campaign itself, and the pre-test function allows for a determination of exactly how much knowledge was gained or lost over the course of the campaign, this particular design also has drawbacks. First, because this design requires data collection at two points in time rather than one point in time, it is less economical and less efficient than the post-test only control group design. The process of giving participants the same test twice also raises concerns about the validity of results. It may be argued that scores may differ from time one to time two simply because participants have taken the test before and they may have discussed some answers or even looked up the answers to some of the questions they did not know. This effect should hold constant across treatment and control conditions, so increases in scores due to this effect should occur equally in both conditions, making any further differences attributable to the campaign.

Another concern may be that giving students the pre-test will prime them to pay closer attention to the campaign being conducted around them than they would have if they had not been tested on the material. For example, a student who did not know the answer to a particular test question, who then one week later sees a bulletin board with the answer to the question posted on it, may be more likely to retain that information from the bulletin board than he or she would have had it not been a test question previously missed. If this is the case, then the results of the post-test may make the campaign messages appear more effective than they actually would be on high school students who had not been pre-tested.

3. **Solomon Four Group Design**: The final true experimental design eliminates both concerns from the post-test only control group design about the limitations of random assignment in equalizing scores across conditions, as well as the possibility of the previously discussed priming effect in the pre-test post-test control group. As suggested by its name, the Solomon Four group design divides participants into four groups, two of which are used to conduct a post-test only control group design, and two of which are used to conduct a pre-test post-test control group design. Thus, the four groups composing the Solomon Four are the control group for the post-test only control group design, the treatment group for the post-test only control group design, the control group for the pre-test post-test control group design, and the treatment group for the pre-test post-test control group design.

Theoretically, a Solomon Four group design may be considered "the best of both worlds." Comparing the results from the pre-test post-test control group design to the post-test only control group design can eliminate concerns about trusting too much in random assignment to equalize groups as well as concerns about testing effects from pre-test to post-test. However, because this design involves twice as many groups, it is more costly and complicated than other true experimental designs. For this reason, a Solomon Four group design is rarely used in experimental research.

While the post-test only control group design and the pre-test post-test control group design are less costly and less complex than the Solomon Four group design, they are still often an unrealistic expectation for practitioners with limited time and resources. In the field of emergency management, it is also often true that practitioners cannot withhold information from a certain segment of the population, such as those who are selected to serve as an experimental control group. Therefore, many communication campaigns choose to use a *quasi-experimental* design instead. Quasi-experiments are similar to true experiments in most respects, other than they do not fulfill all true experiment design requirements (usually because they do not assume random assignment to treatment and control conditions). Only true experimental designs allow for the attribution of effects to the campaign and not some other variable. However, under those conditions in which true experiments are not possible, quasi-experiments can still provide information about the target audience regarding the campaign's dependent variable.

Three types of commonly used quasi-experimental designs are discussed in the following.

1. **One-Shot Case Study**: A one-shot case study involves a single group of people who receive a treatment and are assessed in terms of the outcome variable only after the conclusion of the campaign. Because this design provides a treatment to only one group at a single point in time, there is no point of comparison to detect what types of effects were produced by the campaign. For example, even if, at the conclusion of the campaign, the target audience appears to be taking the actions recommended by the campaign, practitioners cannot be sure that they are doing so in response to the campaign or if they had already engaged in the behavior prior to the campaign. For this reason, the one-shot case study is not recommended as a proper evaluation of campaign effects.
2. **Non-Equivalent Groups Design**: The non-equivalent groups design is similar to the post-test only control group design except that rather than randomly assigning all participants to either a treatment or control group, the practitioner uses two intact groups to serve as the experimental and control groups. For example, in the previous scenario involving educating high school students about emergency preparedness, a non-equivalent groups design might simply designate two schools that had

similar characteristics (e.g., private schools from small rural towns) to serve as the control group and the treatment group and then analyze the data at the individual level. Because this process did not involve random assignment to conditions, and there was no pre-test provided, practitioners cannot know how the groups differed in terms of the dependent variable before the treatment was provided. Therefore, it would be inappropriate to interpret the difference between knowledge scores in the treatment group and the control group as campaign effect size. However, because there is no theoretical reason to suspect that the groups may differ in scores, an outcome assessment indicating that knowledge scores in the group that received the treatment were considerably higher than knowledge scores in the group that received no treatment may reassure researchers that the campaign produced the intended effect better than any outcome that may be provided by the one-shot case study.

3. **One-Group Pre-Test Post-Test Design**: The one group pre-test post-test design is similar to the pre-test post-test control group design, except it does not involve a control group. If practitioners are forced to use a quasi-experimental design due to time, monetary, or logistic constraints, this design is probably the best option for gaining insight into campaign effectiveness. However, because there is no control group for comparison, it is not appropriate to attribute changes from pre-test to post-test to the campaign, as one cannot be sure that changes were not due to testing effects or some other incidental effect that occurred at the same time as the campaign.

When drawing conclusions about campaign effects, it is important to remember to associate the effects achieved or lack thereof with the campaign that actually took place, as opposed to the campaign that was planned. This is another reason why process evaluation is essential. If a major channel of message dissemination is in brochures that never actually are distributed to the population of interest, and the final assessment of the campaign indicates no change from pre-test to post-test in terms of the outcome variable, then it would be incorrect to conclude that the brochures themselves were ineffective. If people are not exposed to the message, researchers cannot use post-test measurements to make judgments about the effectiveness of the message.

The point of evaluating the effectiveness of a campaign is to provide insight into what has worked and what has not worked in meeting the preset campaign objectives, so that future attempts may use this information in guiding campaign design. Thus, it is important to keep records of all aspects of the campaign, including all background research, focus group and other formative research information, materials that were pretested and those that were distributed, pre-test measures, process evaluation information, and post-test measures. There is no guarantee that even the most carefully designed campaign will meet its intended objectives. Keeping clean records of all campaign materials ensures that something can be learned from the process, regardless of the outcome. Documenting all procedures and their impact will also help to save money in future campaigns by providing insight into what has and has not worked with regard to the population or the behavior of interest in the past. If practitioners hired professionals to conduct evaluations, they may want to share their results with others, which could include publishing them. Publishing results helps to increase campaign success and decrease setbacks in emergency preparedness knowledge and behavior by contributing to the knowledge base about how to effectively communicate risks and promote behavior change among different segments of the lay population.

CONCLUSION

The purpose of this chapter was to provide guidance for campaign implementation and evaluation. In the initial planning stages, campaign planners must set specific goals and objectives for the campaign to meet. Meeting these pre-set objectives is the entire purpose of the campaign. Therefore, it is crucial that these objectives be measurable in order to determine objectively whether the campaign had its intended effect.

There is always the possibility that a campaign produces *effects* but is not considered *effective* in meeting its intended goals. Sometimes campaigns may even bring about effects that move the target audience further *away* from campaign designers' goals. Process evaluations help to recognize these unintended effects and allow researchers to end or reverse them as quickly as possible.

In addition to preventing future harm, campaign evaluation also forces campaign planners to set measurable objectives that the campaign seeks to achieve. Public disaster preparedness education campaigns generally have three overarching goals:

1. Raising public awareness
2. Guiding public behavior
3. Warning the public

Objectives are specific measurable outcomes derived from these goals. Campaign success is determined solely by the degree to which it meets the pre-set objectives, rather than sheer exposure or the degree to which people "like" or "enjoy" the materials.

The authors of this book strongly recommend that campaign planners hire professionals to assess the true outcomes of their campaign. This requires funding that will draw from limited budgets, but it can result in significant positive dividends by preventing counterproductive effects or by providing a template for future campaigns.

Using a true experimental design is the only way to determine conclusively that effects observed in evaluation are due to the campaign and not some outside factor. This type of design requires a treatment group, a control group, and random assignment to conditions. This chapter discussed three types of true experimental designs:

1. Post-test only control group design
2. Pre-test post-test control group design
3. Solomon Four group design

Sometimes resources do not allow for true experimental designs. In this case, quasi-experimental designs can provide information about the target audience. Quasi-experiments are similar to experiments in most respects, but they fall short of fulfilling the requirements of true experimental designs, usually because they do not assume random assignment to treatment and control conditions. This chapter discussed three different types of quasi-experimental designs:

1. One-shot case study
2. Non-equivalent groups design
3. One group pre-test, post-test design

It also issued caution regarding the drawing of conclusions from each.

When drawing conclusions about a campaign, it is also important to attribute effects to the campaign that took place, rather than the campaign that was planned. The point of evaluating the effectiveness of a campaign is to provide insight into what has worked and what has not worked in meeting pre-set campaign objectives, so that campaign designers can learn from and apply this information in future campaign designs.

REFERENCES

Campbell, D. T., and J. C. Stanley. 1963. "Experimental and Quasi-Experimental Designs for Research on Teaching." In N. L. Gage (Ed.), *Handbook of Research on Teaching*. Chicago: Rand McNally.

Salmon, C. T. and L. Murray-Johnson. 2001. Communication campaign effectiveness: Critical distinctions. In R. E. Rice and C. K. Atkin (Eds.), *Public Communication Campaigns, Third Edition*. Thousand Oaks, CA: Sage.

Werch, C. E., and D. M. Owen. 2002. Iatrogenic effects of alcohol and drug prevention programs. *Journal of Studies on Alcohol*, 63, 581–590.

6

Program Support

INTRODUCTION

All public education endeavors require financial support appropriate for the levels of effort and exposure anticipated. If the campaign is to be ongoing, that funding must be sustainable. While there are projects that are developed upon a grant or budget that is sufficient to support the project in full, most will require a concerted effort to secure additional resources from elsewhere.

By the time most project planners with small or insufficient coffers have begun thinking about funding shortfalls and budgetary constraints, they have likely completed the first step in the fundraising process — namely, *recognizing and accepting that their organization needs additional funds to either begin a new project or continue an existing one.*

Most projects, even those that are based upon well-established ideas and impressions, or which have a fully developed project plan, must raise or locate additional support in the form of money, in-kind assistance, or labor. The solution to this problem comes with knowing how and where to acquire such resources. A fundraising plan is an effective tool to guide that effort.

In order to design and run a fundraising campaign, it is first necessary to determine exactly how much money and what resources are needed. These figures are known as the fundraising goal, or *the amount upon which all components of the fundraising campaign are based.* Accuracy in determining this number is vital, as it immediately becomes the motivating factor for the fundraising team and the primary performance measure in determining whether your organization is successful in its efforts.

In determining the fundraising goal, it is important that you do not set your organization up for failure. The goal should be based upon two determining factors:

1. How much money is actually needed
2. How much money can realistically be raised (see Sidebar 6.1)

The amount of money that is actually needed is generally easy to calculate or estimate. Planners should have an accurate impression or assessment of the organization supporting the project, including its needs and expenses. Planners should consider at length the specific needs of their project. If a project budget has not been developed already, it is a good time to do so. Project budgets should include all possible costs that may be incurred across the life of the project, including equipment, rental space, utilities (new phone lines, for instance), services, and any other fees. The fundraising process itself should be included if there are associated expenses. If this project is to be ongoing, planners must be sure to calculate ongoing/recurrent fees and costs. Planners do not want to reach their fundraising goal only to find that they are still unable to implement their project because some unforeseen or ill-considered expenses were omitted from the initial analysis. For this reason, it is good to include some leeway in the budget that accommodates most unknown expenses when such uncertainty exists. A budget is considered the "road map" to the fundraising goal, so detail counts.

For planners embarking upon a fundraising effort for the first time, it may be difficult to gauge the potential of the fundraising effort, and forecasts are based upon assumptions rather than experience. When grants or program funds are the target of the fundraising effort, there usually exists a general range of funding that can be expected. However, when planners approach the community or beyond in order to acquire resources, they may need to consider the abilities of residents, businesses, and other stakeholders to provide such funding or other resources, and the salience of the issue being addressed by the preparedness effort. Most emergency management projects appeal to a wide range of audiences, as the products of such efforts ultimately build tangible community strength.

Sidebar 6.1: Setting Fundraising Goals

If possible, fundraising potential can be measured by assessing previous community support for other efforts or organizations. This can be done by contacting partner organizations or other network contacts. Tony Poderis, a successful fundraising expert who has developed several successful fundraising tools, provides those tools to groups new to fundraising to help them determine their fundraising goals. The highlights of his findings are listed in the following:

1. A goal must be based upon the ability to raise the money to pay for it, not by deciding how much money is needed to be raised based on the expense. It is vitally important not to let "the tail wag the dog."
2. The amount of money to be raised must meet with the consensus of the organization's volunteer and professional leadership. You cannot have "It's too much" or "It's not enough" divisive arguments within the organization at the time prospects are being solicited.
3. The goal should be enlarged, if at all possible, with foresight and planning to meet future capital needs. This will help to avoid subsequent closely recurring additional capital campaigns, which could stretch the resources of an organization, antagonize prospects, and possibly have a negative effect on annual fund campaigns.
4. The final goal amount must be related to viable potential prospects identified and individually rated for their maximum giving potential and by factoring the leadership's commitment to personally give to the project and raise money for it.
5. The total goal is composed of trustee, individual, corporate, and foundation divisional goals as determined by realistic and appropriate evaluations of their respective potential. Those divisional goals must never be arbitrarily set.
6. The fundraising goal is more than just establishing the final number. The organization should determine other goals in order to have certain amounts of cash at various stages of the project's development to pay ongoing expenses. It is important to solicit prospects capable of providing early and upfront cash to help meet those developing money needs.

7. The amount to be raised should be influenced by advance leadership pledges, suggested to represent approximately one-quarter of the goal as determined by some form of a pre-campaign feasibility effort. As much as possible, the board of trustees' aggregate contribution should represent at least one-third of the total goal.
8. The goal should be related to the fewest number of gifts in the largest possible amounts: one-third of the money should be raised from about 15 gifts, the next one-third from an additional 75 to 100 gifts, and the last one-third from all other gifts.
9. Non-cash contributions, such as in-kind goods, products, and services, should be factored into the goal as much as possible to help lessen the need for actual cash.
10. Major benefactors recognize that it costs money to raise money. They will support a capital campaign goal that incorporates all reasonable campaign fundraising expenditures, including professional fundraising consultants' fees. While such campaigns are not conducted by employing professional consultants whose fees are initially based upon a percentage of the goal or actual funds raised, it generally works out that total expenses will be in the 5 to 8% range relative to the goal.

Source: Poderis, Tony. 1997. 12 Things You Should Know About Setting A Capital Campaign Goal. Fund Raising Forum. http://www.raise-funds.com/998forum.html.

Your available fundraising resources are the reality against which your goal should be measured for feasibility. These resources consist of fundraising campaign leadership and solicitors available to work a campaign, and a realistic, evaluated list of prospective donors. If the resources are insufficient to raise the money you have targeted in your goal, there are only two available options:

1. The resources must be enlarged to meet a goal equal to the need.
2. The capital project's expense budget must be reduced to allow the goal to be set lower, at a level consistent with available resources.

Your fundraising campaign must not be set in motion until one or the other of these two options is established.

TYPES OF PROGRAM SUPPORT

There are three major categories of program support that may be acquired:

1. Cash
2. In-kind donations
3. Volunteer resources

Each of these serves a different purpose, and in the case of most projects, all three are needed to some degree.

Cash is the most versatile, as it generally presents few restrictions with regard to when it must be used and for what. Of course, it must be used in support of the project, as dictated in the budget that was used to raise the cash in the first place, but this gives planners a considerable amount of leeway.

In-kind resources, which can come in the form of equipment, supplies, property or office space, and services, can be easier to acquire than cash, as they generally require a smaller financial commitment from donors (see Figure 6.1). Additionally, in-kind resources face far less resource competition internal to the donor organization than does cash.

FIGURE 6.1 In-kind donations can reduce project costs significantly. In this community preparedness day fair, Arlington County, Virginia, firefighters demonstrate victim extraction on a donated automobile (Coppola, 2008).

For example, consider a project that will require the acquisition of 10 computer terminals. For a small organization, it may be difficult or even unrealistic to fundraise from a community whose stakeholders have competing needs for their cash resources to purchase 10 new computers. Even if grants have been awarded for the project, restrictions may limit the uses of the funding provided such that the purchase of computers cannot occur. However, individuals and organizations within the community may have unused computers that they would be willing to lend or donate — a gesture that would likely cost them nothing but give them a tax deduction in return. For the project team, this would provide the same value to the organization — for zero cash outlay — as purchasing new computers with cash.

Volunteer resources are the third program support category. While many people may have little or no money to donate, they often have the time to do so. In addition, when the benefits of the program affect them or their community directly, there is a strong goodwill incentive for people to provide their time and talent on a voluntary basis. Volunteers can be used for almost any aspect of project planning and operation, including graphic design, conducting surveys, community outreach — even fundraising. People of all age groups volunteer their time. These individuals may be reached through community newspapers, recreation departments, schools, and other sources.

SOURCES OF SUPPORT

Sources of funding come in myriad forms — as individuals, businesses, organizations, grants, events, and many others. In this section, several of these forms will be identified, but keep in mind that this is merely a sample of the many standard and creative options planners may encounter or develop in their fundraising efforts.

As is true with successful investors, programs are most successful when they diversify their funding pool — drawing from a full range of options to minimize shortfalls. In most general terms, the various sources and means by and through which funding is raised are presented as follows:

1. Individual donors
2. Business and corporate donors
3. Foundations and other grant-making agencies
4. Local, state, county, and federal government grants

5. Religious organizations
6. Civic organizations/civic clubs
7. Fee generation
8. Partnerships

Individual Donors

Individual donors account for the vast majority of charitable and other philanthropic assistance that is given each year in the United States. Professional fundraisers expect that, on average, three-quarters of the money they collect will come from individuals, and in total, around 90% of money given outside of federal grants comes from individuals. Individual donations are generally smaller than what is attained through other sources, but such donations tend to be the most spontaneous and unrestricted. Cultivating individual donors tends to require a smaller financial commitment than many other fundraising methods, and the associated outreach helps to promote your project throughout the community (as well as to cultivate additional contacts, partners, and supporters). If your organization is a 501(c)(3) nonprofit organization, any donations are tax deductible for the donor. Finally, individuals who are secured as donors tend to give again when asked in the future (for even greater amounts if they are treated well the first time around), which makes subsequent fundraising campaigns easier.

Individuals, like all donors, must be identified and approached. One of the great benefits of working with individual donors is that there is an extensive range of options by which this can be done. A 1996 study that was conducted by the Independent Sector (2008), an association of nonprofit organizations, asked individual donors the question, "How important is each of the following reasons to you for contributing to a charitable organization?" Participants listed the following reasons as either "very important" or "somewhat important:"

- 72.1% — someone I know well asked
- 60.7% — have volunteered at the organization
- 59.1% — asked by clergy
- 43.3% — read or heard a news story
- 38.2% — asked at work
- 36.2% — someone came to the door asking me to give
- 29.7% — asked in a telethon/radiothon
- 28.6% — received a letter asking me to give

- 17.1% — read a newspaper or magazine advertisement asking me to give
- 16.9% — saw a television commercial asking me to give

There are several ways in which you can approach individuals to present your project and to ask for support. They include:

- Direct mail
- Special events or activities
- Internet fundraising
- Telephone solicitation
- Door-to-door solicitation
- Planned gifts
- Workplace appeals
- Advertising

Depending upon the size of the fundraising goal, the community makeup, and the resources available to the planning team (for instance, a volunteer or member who can design a fundraising Website), organizations will likely employ a strategy that includes several of these categories.

Concluding Remarks about Individual Donors

For most organizations, individual donors are their greatest resource — to neglect them will severely limit potential. Donors become more than givers of money, they become vested in the organization and harbingers of the cause to which they give. It should be mentioned at this point that there are people in the community who can give much more than the average donor. These people, referred to as *major donors,* require special tactics to approach. Major donors deserve special attention because the time invested in educating them about the organization and the project, specifically about how their donation will make a difference, is both wise and worth the effort. Such donors should be approached by the organization's most senior executives, and deserve as much professionally formatted information as possible (including a full project proposal, if appropriate). Programs have found their entire goal met by the gift of one major donor.

Business and Corporate Donors

Businesses and corporations are second in terms of charitable contributions provided annually. This category of donors includes multi-million and billion dollar mega-corporations whose products and services permeate all levels of society and the corner store on any block in every community. Despite the fact that many large corporations maintain distinct gift-giving foundations, which are the subject of another section, many large companies still provide donations directly through the company itself. The motivation behind each of these business types' giving dictates how they are most effectively approached. The following is a general discussion on business and corporate donors.

The bigger a corporation or business is, naturally, the larger its philanthropic budget will be. However, these larger businesses also typically have more people knocking on their doors for money. Having a personal connection with the company will therefore play a large part in whether a requesting organization receives funds. For instance, if the donor corporation is headquartered or has a satellite office in the target community, has a large sales base there, or even has a board member who happened to have spent their childhood there, funding likelihood is much greater. How an organization seeking funds determines which corporations have some connection to its community does require some creativity and research, but this determination is nonetheless a prerequisite to considering corporations for donations.

An organization approaching a business for program support should consider the interaction a business proposition as much as a philanthropic gesture. Planners should determine what they could offer the corporation in return for its goodwill (especially if it is something that others cannot offer). Fundraisers need to sell their organization, their employees, and their proposed project in order to make the decision to donate seem like a wise one. The challenge is determining what can be offered in return.

The number one thing that all businesses seek in return for philanthropic giving is an improvement in their good reputation. If an organization's standing in the community is high, then association with it will help to lift the company's image. A better reputation often translates to more products sold and, in turn, more revenue. Executives and decision-makers need only to be convinced that this positive return is possible. Fundraisers must investigate the company in order to discover how it operates and, more importantly, how it considers requests for donations.

Using resources that are readily available, such as the phone book, the Internet, Chamber of Commerce directories, corporate giving reports, local business directories, and so forth, planners can make a list of companies whose missions parallel their own. Subscriptions to philanthropic services, such as the Foundation Center (www.fdncenter.org), can provide easy access to this kind of information. In small communities, just about any company could be approached because the project, being disaster preparedness based, would be in the interest of all businesses, as it benefits the community as a whole. Moreover, when the community is a large one, because the number of groups seeking donations is also large, organizations should limit themselves to those companies with which they have the greatest likelihood of success. Generally, fundraising teams make a large list from which they focus on between 10 and 20 companies that are initially approached.

Most of the information regarding a company's philanthropic tendencies can be found on its Website and in its annual report (if one is maintained). If it can be determined how much the corporation or business has given to the type of organization planning the public education project in question, this information will be valuable as well. Organizations do not want to be asking for more than the greatest amount ever provided, or less than the least. If the average gift size works well for the program's needs, then the organization would be appropriate.

Small Businesses

Small businesses exist in every community. These entities tend to have a much greater stake in community success and viability because they are much more dependent upon it for their own survival and success. While they share many of the same goals as the larger corporations, such as the need to turn a profit, there are distinct differences between the two and, as such, there are distinct differences in how they are approached for philanthropic giving.

There are almost no reasons why small businesses should be neglected by a fundraising effort. As long as planners understand that donations are likely to be smaller, donor interests are likely to be focused, and the need for personal relationships is great, they will have a big chance of raising funds or acquiring goods from one or more of the millions of small businesses that operate throughout every corner of the nation.

A national survey of small businesses commissioned by the Better Business Bureau (2001) revealed important data about how small businesses contribute to community causes. The summarized findings of this survey are presented in Sidebar 6.2.

Sidebar 6.2: Small Business Fundraising Survey Results

Almost all small businesses in the United States participate in charitable giving. In fact, 92% of companies were solicited and 91% donated. For new small businesses, 85% have been approached and 82% have given.

Contributions are not limited to cash. Sixty-three percent of small businesses provided in-kind contributions of products or services.

Most (85%) small businesses are willing to participate in fundraising events. The most common event-related support in which small businesses participate includes the purchase of advertisements in charity event programs or publications (60%) and sponsoring youth sports teams or athletic leagues (51%). More than 3 in 10 small businesses also participated in each of the following: charity auctions (37%); fundraising benefits, dinners, or galas (35%); and special fundraising events like walkathons (32%).

Altruistic concerns matter the most to small businesses when selecting donation recipients. Ninety-two percent stated that helping the community or society is key to their selection, and half claim such concerns are the single most important consideration.

Second only to altruistic concerns in determining recipients, listed by 82% of respondents, is the small business owner's preference or recommendation.

Seventy-eight percent of small businesses claim that locally based efforts are preferred, while less than 1% gives to nationally based organizations.

Small businesses rely heavily on the recipient organizations themselves, rather than on outside sources, for information used to make funding decisions.

Most (85%) small businesses give less than $5,000 in philanthropic donations each year.

The top five types of recipient organizations supported by small business are:

Police and firefighter (58%)
Educational (53%)
Religious (50%)
Social services (44%)
Health (41%)

Foundations

Foundations are nonprofit organizations, either associated with or independent of other private or public organizations, that exist almost exclusively for donating money to worthy causes. Foundations in the United States boast assets of almost $400 billion and give away almost $25 billion each year — approximately 10% of all philanthropic funding. Moreover, the amount that foundations are giving rises each year as the capital upon which their interest-based funding grows (with single-year growth rates of up to 22% experienced in the recent past).

All foundations have a mission and program areas that define what types of projects they fund and what types of nonprofits or other agencies they prefer to work with. Without fail, all emergency preparedness public education efforts qualify in both of these regards for many of the *tens of thousands* of foundations that exist in the United States. However, to acquire these resources, planners will need to have at least the following prepared:

1. Good research
2. A great proposal
3. Time, and lots of it, to wait for the approval process to run its course

The key to winning grants from foundations is a good match between the project's mission and goals and the mission and goals of the foundation. Foundations do not hide their interests and intentions, so with good research planners can increase their funding likelihood considerably. Most foundations require a proposal, and there are often strict guidelines about how those proposals should look and what type of project may be funded.

There are four different categories of foundations, each of which includes members of all sizes and missions. They include (1) community, (2) public, (3) family, and (4) private.

Community Foundations

A community foundation is a philanthropic organization, organized and operated primarily as a permanent collection of endowed funds, the earnings of which are used for the long-term benefit of a geographically defined community. A community foundation is tax exempt, incorporated, not-for-profit, organizationally autonomous, and cannot be controlled directly or indirectly by government at any level, corporations, associations and their members, or individuals.

The primary purpose of community foundations is to provide charitable support to their local communities. They do this by building endowments with contributions from local residents, and administering them for the benefit of their communities. They also administer non-endowment funds. In essence, a community foundation is an organization that gives support, primarily in the form of money, to a specific area — a town, city, county, state, region, or country.

Community foundation funds can be restricted not only to a defined physical area, but also to a program area. Oftentimes, however, community foundations have a certain percentage of funds earmarked for specific purposes, while the remaining share is unrestricted and distributed at the discretion of the community foundation management.

Grant application procedures for community foundations can take a long time to process. The foundation's programmatic restrictions, their procedures and deadlines for applying, and any application material should either be on the foundation's Website or obtained from the foundation directly using contact information their Website provides. Grants tend to be made according to award schedules, biannually, quarterly, or at any other interval as they choose.

Public Foundations

Public foundations are foundations that, as defined by the Internal Revenue Service (IRS), receive more than one-third of their funding from the public at large. Many religious, educational, medical, and other population-targeted foundations are considered public foundations if they meet this first criterion. While public foundations often engage in charitable activities (services to the public), grant giving is a significant part of their collective mission.

Family Foundations

Family foundations (often called either large family foundations or small family foundations) are private foundations whose philanthropic funding base is derived from the gifts of a single family. Additionally, family members typically compose at least one seat, if not the majority of the seats, on the foundation's Board of Directors. Family foundations account for a major portion of foundation giving in the United States, most notably since Bill and Melinda Gates contributed more than $4 billion to the Bill and Melinda Gates Foundation. In the United States, families manage approximately two-thirds of the estimated 40,000 private foundations and make grants totaling more than $7 billion per year ($7 of $22.8 billion total).

Private Foundations

Private foundations are non-governmental, nonprofit organizations that have collected their funding base from either one source or a few sources. Corporations, families, small businesses, and other entities all have and continue to form grant-giving foundations such as these. Six percent of philanthropic giving in the United States comes from private foundations.

Private foundations often operate with much greater leeway in their giving than the public or community foundations, and can move much faster from application to award (see Sidebar 6.3). Private foundations can and do give to the whole range of organizations that exists throughout the nation. These organizations can have self-imposed restrictions, and often do. Whether it is geographic or programmatic, those applying must be aware of any of these restrictions before going through the often-lengthy application process.

These foundations differ from regular corporate and business donations in that they come from a separate entity, established by the corporation or business, whose single goal is philanthropy — remember that this is a nonprofit organization. The structure for giving is established, straightforward, and created primarily to guide potential grantees in applying.

Sidebar 6.3: The Differences between Public and Private Sources of Funding

Advantages — Public Sources

- Purpose set by legislation
- Focus on functions usually affecting significant groups in society
- Have the most money and more likely to award large grants/contracts
- More likely to pay all project costs and cover indirect costs
- Easier to find information about and to stay current on project needs/interests
- Application processes and deadlines are public information and very firm

- Use prescribed formats for proposals — many use "common" application forms
- Possibilities of renewal known up front
- Plentiful staff resources — most projects have specific contact person
- More likely to have resources for technical assistance
- Funds available to a wider array of organizations (for-profit and nonprofit)
- Accountable to elected officials if administrative staff do not follow the rules

Advantages — Private Sources

- More likely to focus on emerging issues, new needs, populations not yet recognized as "special interests"
- Often willing to pool resources with other funders
- Wide range in size of available grants — some can make very large awards, others are strictly for small, local projects
- More willing source of start-up or experimental funds
- Full-length, complex proposals not always necessary
- Can be much more flexible in responding to unique needs and circumstances
- Able to avoid bureaucratic requirements for administering grants
- Can often provide alternative forms of assistance, i.e., software/hardware donations, materials, expertise, etc.
- Fewer applicants in most cases
- Can generally be much more informal and willing to help with the proposal process

Disadvantages — Public Sources

- Much more bureaucratic
- Lengthy proposal requirements and complex application, administration, and compliance procedures
- Often require institutional cost sharing and matching
- Reviewers tend to favor established applicants
- Sometimes difficult to sell new ideas and high-risk approaches
- Cost to applicants much higher — expensive application and compliance procedures
- Changing political trends affect security of some programs — availability of funds can change rapidly

Disadvantages — Private Sources

- Average grant size usually much smaller
- Priorities can change rapidly, continuation of support can be difficult to predict
- Applicants have limited influence on the decision-making process
- Information on policies and procedures must be researched and can be time consuming
- Less likely to cover all project costs and most do not cover indirect costs
- Limited staff — fewer opportunities for personal contact and site visits
- May not be clear about reasons for rejection — hard to improve for second attempt

Source: Hall, Mary. 1988. *Getting Funded: A Complete Guide to Proposal Writing,* 3rd ed. Portland, OR: Continuing Education Publications.

Finding Appropriate Foundations

To find appropriate matches between your organization and foundations with the money sought, planners need to define their organization according to its mission. Then, the same needs to be done for their project. Planners should not limit themselves by getting too specific — it is preferable to make a list beginning with the most specific but expanding to the most general.

For example, imagine a project where volunteers will hold presentations in the local community center to teach senior citizens how to prepare for emergencies. The list might look something like this:

- Senior citizens/the elderly
- Emergency preparedness
- Disaster management
- Public safety
- Public education
- Training
- Homeland security
- Community development

Matches among the various foundations must now be found. As there are over 45,000 foundations operating in the United States alone, it is usually possible to do so. The most appropriate starting point is with community foundations operating in the state in which the project will operate. Then, the search can be expanded beyond local and state borders, with the understanding that competition for resources expands as well.

Several Web-based philanthropic associations assist organizations seeking foundations to suit their needs, although many of these operate through membership fees. A good example is the Foundation Center (www.fdncenter.org), which allows prospective grantees to search through thousands of foundation Websites according to either geographic area served or the sector in which the grant applies (public safety, for example). The following link can take you to the free sector search engine, but other search methods require subscription: http://lnp.foundationcenter.org/finder.html

Local, State, County, and Federal Government Grants

While not all government agencies make grants, there are agencies at all *levels* of government that do make them, for example, the local, state, and federal levels. These grants are a perpetual but unpredictable source of funding for organizations of all types. It is understood that finding government grants can be an incredibly challenging task. An open mind, vigilance (in watching for grant programs as they appear), and persistence are the keys to getting funding through these sources. Having a grasp of the methods by which the federal, state, and local government agencies announce their grants is a great way to stay "in the loop." Unfortunately, this is usually just one part of a multi-faceted approach that is required if you are to take advantage of all opportunities for which your organization qualifies.

Like the foundations, government agencies give grants according to specific program areas in line with their stated mission — this is increasingly true as you rise from the local to national level. Though not without exception, very rarely is government funding indiscriminant. For instance, the Department of Health and Human Services is likely to give grants only for health-related programs, while the Department of Homeland Security will focus on public safety, emergency preparedness, terrorism prevention, and immigration issues. If you can think creatively about your project, looking at it from as many programmatic vantage points as possible (as was mentioned earlier in reference to the foundations), you might find that your program is eligible for grants you had not even considered because you did not think of your program in the right context.

Government agencies are strict in awarding and disbursing of funding. They require that applying organizations follow pre-set guidelines and submit applications by pre-set deadlines. These guidelines, however, are well defined and easily accessible from the agencies themselves or through the many online search engines dedicated to government grants. There are even public-oriented publications distributed solely for letting prospective grantees know about the existence of grants as they are announced. Keep in mind that at each successively lower level of government, these processes tend to be more lax, and the turnaround times for grants tend to be much shorter, but the mechanisms for announcing the grants can be less effective.

For NGOs, eligibility requirements will be the greatest obstacle. Many of the federal and state programs, for instance, are limited to state and local governments, with explicit definitions of both listed in the application announcements. Many nonprofits searching for grants do find programs that seem appropriate to their needs in terms of the programmatic details, but see that the eligibility requirements appear to preclude them and they move on. What they are not aware of is that *partnerships* can give them the eligibility they seek (see the "Partnerships" section later in this chapter for more information).

Local Government

Cities and towns control their own budgets, and decide where to spend their money. Unfortunately, very few towns have a lot left over to create formalized grant programs. However, special projects that fall outside of any formalized grant programs are much more common at this level of government than any other, and as both a member of the community and an organization whose goal centers upon improving life in the community, organizations may be well placed to get a hold on any resources that exist.

Planners will need to do some firsthand research to find out if there is any funding available through their municipal government. The best and likely only chances come when planners make personal contact with leading local government officials themselves (see Sidebar 6.4). If a town is very small, this might involve setting up a meeting with the mayor or other ranking elected official, while in larger communities you may have to weed through a bureaucratic puzzle to discover what government agency manages the issues your organization or project intends to address.

Once contact has been made, it is up to the fundraisers to accurately explain their organization and project, and how their project will immediately help the community. Each day, people and organizations approach the local government looking for funding, so you must be prepared and able to convince whomever you are speaking with that your project is the worthy one. Having a high-quality proposal will help your case immeasurably, especially if given only a short amount of time to make a case.

Although the local government official first approached may not be able to offer funding, there is a very good chance that he or she will have a handful of ideas to help you find the money you need. By involving the mayor or the local government as a partner in the project, there will be an even greater chance of acquiring funding. The same could be said for the police chief of fire chief. In the past, organizations have even gone so far as to associate themselves through their title, such as the "Springfield Mayor's Prepared Corps."

Sidebar 6.4: Directories of Local Government Websites

- **American hometowns — cities, counties, and towns** — Click on one of the 50 state directories to get to local governments. http://www.usa.gov/Agencies/Local_Government/Cities.shtml
- **Directory of county, city, and other local government Websites** — Click on your state to find a comprehensive directory of county, city, regional, and other local Websites. http://www.statelocalgov.net/
- **FirstGov gateway for local and state governments** — Information and services for state and local government employees such as grants, disaster management, and best practices. http://www.usa.gov/Government/State_Local.shtml
- **International city/county management association** — Professional and educational organization for members of the local government community. http://www2.icma.org/main/sc.asp?t=0
- **Locate in-person service centers in your community** — Find government offices and services such as local motor vehicle offices, post offices, and nursing home facilities. http://www.usa.gov/Citizen/Find_Services.shtml

- **National Association of Counties** — Professional association of county officials that promotes public understanding of U.S. counties. http://www.naco.org/
- **National League of Cities** — Largest national organization of municipal governments, representing 18,000 cities, villages, and towns in the United States http://www.nlc.org
- **Summary information for U.S. counties** — Information about county officials, county seats, cities in a county, county statistics, and more. http://www.naco.org/Template.cfm?Section=Find_a_County&Template=/cffiles/counties/usamap.cfm
- **U.S. conference of mayors** — Official nonpartisan organization of U.S. mayors, representing cities with populations of 30,000 or more. http://www.usmayors.org/meetmayors/citiesonline.asp
- **U.S. mayors** — Locate mayors by name, city, or population size. http://www.usmayors.org/USCM/cgi-bin/database_search4.asp

Your phone book should also maintain a blue pages section that lists all local government contacts. Libraries commonly have this information as well, and library staff can be a great resource.

State and County Governments

State and county governments are the second option in the search for government funding (see Sidebar 6.5). Oftentimes, the federal government disburses block grants or other formula grants to state governments to distribute to the cities and towns as they see fit. Other times, the state and county governments themselves set aside state-generated funding and create grant programs for which communities and organizations can compete. Either way, planners will need to talk to the appropriate offices to find out if there is any money for which their organization may qualify.

As the projects discussed in this book are emergency- or disaster management-based, it is most appropriate that the first agencies contacted at the state government level be their Office of Homeland Security (which all states now maintain), and the Office of Emergency Management or Public Safety. Then, depending upon the project scope, it may be realistic to talk to representatives in any of the following offices (variations on these names are used in each state and county):

- Office of the Governor
- County Commissioners Office
- Department of Education
- Commission on Aging/Elderly
- Fire Prevention/Control Commission
- Offices of Protection of /Advocacy for Persons with Disabilities
- Office of Emergency Management/Civil Protection/Public Safety
- Office of Homeland Security
- Public Health

Sidebar 6.5: Directories of State or County Websites

State Offices of Homeland Security: http://www.dhs.gov/xgovt/editorial_0291.shtm or http://www.ncsl.org/programs/legman/nlssa/sthomelandoffcs.htm
State Offices of Emergency Management: http://www.fema.gov/about/contact/statedr.shtm
State Government Websites: http://www.usa.gov/Agencies/State_and_Territories.shtml
County Websites: http://www.naco.org/Template.cfm?Section=Find_a_County&Template=/cffiles/counties/usamap.cfm

---

The Federal Government

The federal government, naturally, offers the greatest total amount of funding of all the government levels. The key factor in being awarded a federal grant is knowing that it exists. Once aware of the grant, and having determined that their organization type is eligible, planners can begin crafting a proposal that will place their project or program within the restrictions and goal of the grant program. New federal government regulations have been set that require grant announcement and application documents to *all* be posted online, giving you all the tools you will need wherever you are.

Fundraisers can use two primary tools to research federal grants. They are

1. The Catalog of Federal Domestic Assistance
2. The Grants.gov Website

The Catalog of Federal Domestic Assistance (CFDA) — The online Catalog of Federal Domestic Assistance provides access to a database of all federal programs available to state and local governments (including the District of Columbia); federally recognized Indian tribal governments; territories (and possessions) of the United States; domestic public, quasi-public, and private profit and nonprofit organizations and institutions; specialized groups; and individuals. After finding an appropriate program, applicants can contact the administering office and find out how to apply. The URL for the CFDA is http://www.cfda.gov/

Grants.gov — The charter of Grants.gov, one of 24 "President's Management Agenda E-Government initiatives," is to provide a simple, unified electronic storefront for interactions between grant applicants and the federal agencies that manage grant funds. There are 26 federal grant-making agencies and more than 900 individual grant programs that award more than $350 billion in grants each year. All grant seekers, which includes state, local, and tribal governments, academia and research institutions, and not-for-profits, can visit this Website to access grant funds available across the federal government. The URL for Grants.gov is www.grants.gov.

Grants.gov (see Sidebar 6.6) also sends out emails to prospective grantees as grants become available, minimizing the chance that they miss out on an opportunity. Those interested can register for this service by visiting the Grant Opportunities email site on the Grants.gov Website at: http://www.grants.gov/applicants/email_subscription.jsp. Be sure to see the resources section for links to recommended federal grant programs related to emergency management issues.

Sidebar 6.6: Grants.gov

In short, Grants.gov provides:

- A single source for finding grant opportunities
- A standardized manner of locating and learning more about funding opportunities
- A single, secure, and reliable source for applying for federal grants online
- A simplified grant application process with reduction of paperwork
- A unified interface for all agencies to announce their grant opportunities, and for all grant applicants to find and apply for those opportunities

The 900+ grant programs listed on this Website fall into 21 categories defined by the Catalog of Federal Domestic Assistance, including:

- Agriculture
- Arts
- Business and commerce
- Community development
- Consumer protection
- Disaster prevention and relief
- Education
- Employment, labor, and training
- Energy
- Environmental quality
- Food and nutrition
- Health
- Housing
- Humanities
- Information and statistics
- Law, justice, and legal services
- Natural resources
- Regional development
- Science and technology
- Social services and income security
- Transportation

The 26 agencies that post grants on the Website include:

- Agency for International Development http://www.usaid.gov/
- Corporation for Community Service http://www.nationalservice.org/
- Department of Agriculture http://www.usda.gov/
- Department of Commerce http://www.doc.gov/
- Department of Defense http://www.defense.gov/
- Department of Education http://www.ed.gov/
- Department of Energy http://www.doe.gov/
- Department of Health and Human Services http://www.hhs.gov/
- Department of Housing and Urban Development http://www.hud.gov/
- Department of the Interior http://www.doi.gov/
- Department of Justice http://www.usdoj.gov/

- Department of Labor http://www.dol.gov/
- Department of State http://www.state.gov/
- Department of Transportation http://www.dot.gov/
- Department of the Treasury http://www.ustreas.gov/
- Department of Veterans Affairs http://www.va.gov/
- Environmental Protection Agency http://www.epa.gov/
- Federal Emergency Management Agency http://www.fema.gov/
- Institute for Museum and Library Sciences http://www.imls.gov/
- National Aeronautics and Space Administration http://www.nasa.gov/
- National Archives http://www.archives.gov/
- National Endowment for the Arts http://www.nea.gov/
- National Endowment for the Humanities http://www.neh.gov/
- National Science Foundation http://www.nsf.gov/
- Small Business Administration http://www.sba.gov/
- Social Security Administration http://www.ssa.gov/

Religious Organizations

Every community, no matter how small, is served by one or more religions organizations, each of which likely gives at least a small amount of support back to the community. Every one of these organizations, except in the rarest exceptions, also has a national office that is dedicated to philanthropic activities at the national or international levels. No matter the organization's background, religions organizations are a funding option to be at least considered in any fundraising campaign.

At the national level, many of these religious organizations operate much like the large foundations do — they have a defined competitive process by which applicants submit proposals and grants are awarded. The restrictions imposed by these organizations, however, can be very specific, and, because of these restrictions, the national offices are often an unrealistic source of funding. Additionally, some nonprofits themselves have restrictions that limit their access to religiously based grant-giving organizations. Like all the options listed in this guide, planners will need to see for themselves whether these restrictions exist within their own organization.

However, local branches of these religions, though not as well funded, are often great sources of funding and volunteer assistance. If a project plans to work with disadvantaged groups, its chances of garnering support from this group of funders is even greater. Churches can provide the venue planners may need for events, and have the necessary community support they need to bring in prospective individual donors. As has been said throughout this section, creativity is the major limiting factor when it comes to religious groups as donors.

Civic Organizations and Service Clubs

Civic organization and service clubs such as Kiwanis, the Elks, Rotary, Jaycees, and the Lions are a great contact and funding prospect in local and regional campaigns. These organizations often make grants to organizations in the communities where they operate, and can be a great source of introduction to other funders in the community.

Civic organizations and social groups are perpetually seeking speakers and guests at their meetings and events. Organizations seeking funding can develop a public relations team that works to get on the agenda of those luncheons and meetings to talk about emergency preparedness and, more importantly, to talk about the project being developed. What is gained in return for speaking is the opportunity to meet local business representatives who are able to assist with donations and in-kind contributions, and who might be interested in helping the organization as a technical assistant or in another capacity. For organizations with a board of directors, this method can be useful for finding new board members who, as individuals, are proven donors — and, as representatives from the organizations where they work, they may even be able to provide equipment, publicity, or a venue for a fundraising event.

Fee Generation

If the organization is permitted to do so, fee generation can be a way to ensure small but steady income to keep a program running. This is where an organization offers a product to lure donors, and the profits from those sales entirely go to operational expenses. Many organizations have been very creative in this area, selling both products and services to support their programs. Services offered as part of a project, such as tuition fees and equipment, can be accompanied by a suggested donation fee that benefactors can choose to pay depending upon their resources. Planners may find that, with a good fee generation plan and effective marketing, this form of fundraising alone can serve to offer their project the sustainability for which all organizations wish.

However, fee generation requires significant creativity. For instance, people like to advertise their philanthropic support, which can be accomplished by selling T-shirts that are inexpensive to produce and can give 100% in return for the investment. Actual project participants tend to be especially interested in buying such items. Books or other resources related to the project, such as emergency supply kits, first aid kits, or other preparedness materials, can be sold to generate fees, and people may be more inclined to buy these products if they are informed that they are contributing to a program serving their community. An added benefit of this type of fundraising is that it is a form of financial empowerment that can make an organization feel like it is self-supporting and not entirely dependent upon the handouts of others.

Partnerships

Emergency preparedness and response is a concern of all community stakeholders, whether they are a governmental agency, a religious group, a private business, a school, a hospital, a nonprofit organization, or one of the many other groups that are joined in a civil society. While each of these stakeholders possesses unique resources and skills, each also has unique vulnerabilities and needs. By working in partnership with one or more of the various stakeholders in your community, it is possible to pool together skills and resources while addressing each other's needs and vulnerabilities — the result being increased strength and opportunity for all partners involved.

Partnerships may also provide you access to local, county, state, and federal funding to which planners might otherwise not have access. For instance, FEMA's United States Fire Administration (USFA) provides hundreds of millions of dollars in funding through its Assistance to Firefighters Grant program — more than $600 million this year alone. While only recognized fire departments are eligible for the program, a nonprofit working in partnership with a fire department on a special project, such as assisting in home fire safety audits, could benefit from the grant program despite issues with eligibility. Keeping this in mind as they peruse the long list of federal grants available at any given time, planners need to consider what kind of partnership would give them the eligibility they need, and then approach that partner with a plan or proposal. The more legwork planners offer their partners in the project (many of these first-responder agencies, including fire, police, EMS, and emergency management, have very little room for additional responsibilities), the greater their chances are of developing that partnership.

IMPLEMENTING A FUNDRAISING STRATEGY

Planners should always begin with the sources that offer them the greatest chance for success, and move methodically toward those that offer less confidence. Most fundraisers will say that one should always start with family, friends, employees, volunteers, vendors, a board if one exists, and anyone else that they could say that they know. The unfortunate truth is that as one moves farther and farther away from one's organization and circle of contacts, the more time and effort that will be needed to convince the donor to give. The good news is that closer does not necessarily mean less money.

If someone who is unfamiliar with the organization is approached, one of his or her first inquiries may be about what other sources of funding already exist. They will probably look to see if those with a stake in the organization have come forward with support. If a board exists, for example, and the board has not supported the project, they will wonder why they should. As for staff, the goal is not to raise a great deal of funds but rather to build a high percentage of participation at any level. This show of support can go a long way in convincing other prospects about the dedication of the organization and its people.

Vendors of an organization — those with whom the organization does business — could also be asked to participate in a fundraising project. While some may have policies against giving to organizations with which they do business, it never hurts to ask. Sometimes contacting a vendor can lead to a gift from a larger corporate foundation. Other times it may lead to a reduction in price for the services or products you are already receiving from the vendor. Either way, more funds will be available for the organization's mission.

Local business and community leaders have the next closest stake in the project, and should be contacted next. Planners should present the value of their work in the community and be prepared to discuss ways that they can help publicize the generosity of local businesses. While philanthropy is a primary motivation, businesses and politicians are also pleased to have opportunities for good public relations.

Finally, foundations should be approached with grant proposals. Planners can start with local foundations before moving on to the national ones. It is important to pay attention to guidelines and deadlines, provide what is requested, make personal contact when possible, form relationships that can help a funding decision, and be truthful. Substance, commitment, and conviction can actually outweigh a professionally polished proposal.

No matter the fundraising project, planners should always be prepared to prove that they have done all they can with their local resources before they seek additional outside support.

FUNDRAISING STRATEGY FACTORS

The following factors should be considered as planners create their fundraising strategy. Collectively, these factors are the key determinant variables dictating actions.

- Timing. What is your total time frame in which you need to raise money? Urgency for requirement of funds should be considered. If cash is not needed immediately, you may have the option of asking for pledges.
- Amount of funding required. The size of expected donation should be relative to the size of the project. Ask for a specific dollar amount and be realistic.
- Available resources. Number of volunteers, costs involved, and time (work hours) available to conduct a fundraising campaign.
- Profile of project. In many cases, your donors must be offered tangible and intangible returns (e.g., public relations or direct individual benefits). Identifying the profile of the project will help to determine geographic boundaries of campaign.
- Environment. Availability of funding from some sources may be conditional on the economy.
- Sequence. If more than one source is being considered (e.g., event fundraising used in conjunction with another source), individual contributions may be affected by prior "donations" through special events fundraising. Make sure that your sequence is both logical and set to optimize your chances of being funded.
- Values. The organization must be supportive of the method being used to raise funds. Some types of fundraising activities may not

meet with the approval of the organization or community (e.g., lotteries, sponsorship by certain companies).
- Legal and regulatory issues (see Sidebar 6.7).

Sidebar 6.7: Legal and Regulatory Issues of Fundraising

There are several factors that organizations need to consider to ensure they are legally covered before you begin asking for contributions for your organization or project. The following information was compiled by Idealist.org, an online forum for nonprofit organizations.

Are my fundraising activities appropriate, allowed, and/or legal? Be sure that the money you are collecting is money that you are allowed to use. Many agencies have legal restrictions that limit certain types of private donations, and many states have laws governing giving. You do not want to appear incompetent later on if you are faced with having to return donated money.

Are there any laws or regulations I need to be sure I follow? Many states have what are called *charitable solicitations laws*. Some cities and counties have similar regulations, and it is often hard to find out exactly how they might apply to your plans. There are requirements for one-time or annual registration, and annual reporting. A group of states has come together through the nonprofit sector and developed a uniform registration form, with about 47 participating. For annual financial reporting by charities, many states now use the IRS's Form 990, supplemented by their own form for additional questions. Even to obtain the appropriate forms and instructions from all the appropriate state offices is challenging and time consuming. In many states, out-of-state charities soliciting in the state are required to register. In addition, charities often have to file with the secretary of state as corporations doing business in that state. Some jurisdictions flatly forbid raising money for a charitable organization without an agreement signed in advance with the recipient organization.

The state rules are summarized at http://www.multistatefiling.org. This Website was created to help big organizations that ask for donations in many states, but the state-by-state index of regulations will give you a sense of whether you will need to talk with your local charity-oversight officials before starting your campaign. You can find out if your state accepts the common filing forms at http://www.multistatefiling.org/#yes_states.

Are there any taxes that I might have to pay? You will want to be careful not to mix up the funds you raise in your campaign with your own money. You will need to keep careful records. You might want to open a separate bank account that will be used only for this campaign. Since you are raising the money to pass along to someone else, it should not be considered income that you earned, but you will need to be able to document that you kept the funds completely separate if the IRS or your local tax department should ask.

Are the gifts people give to my campaign tax deductible? That depends. First, it depends on the givers' own tax situation. Careful fundraisers never say, "You can take a tax deduction." What they say (when it is true) is "You may be able to deduct this from your personal income taxes. Consult with your tax advisor or review the IRS regulations that apply to your situation to determine whether a deduction is allowed."

However, tax deductions can only be taken for certain kinds of gifts. If you know your campaign is not about raising money for a tax-deductible purpose, you need to be careful to tell people that right up front.

- The recipient organization has to be recognized by the IRS as a 501(c)(3) — or be an established church. (Churches do not have to register with the government, but they do have to meet certain standards about how they operate. If you have any doubts, ask.)
- Gifts to individuals, no matter how urgent the need, are never tax deductible.
- One result of the rule that gifts to individuals are never deductible is that you must be sure you are raising money for a recognized charity before you ask for any donations. You cannot ask people to give you donations that you will later decide how to use (unless you want to go through the process of setting up a new nonprofit organization, with all the complications in that process, before you start your campaign).

What about enlisting other people to help with my campaign? If you do that, you will need to have some sort of internal controls to be sure that everyone sticks to your standards about how the campaign is run and how the money is handled. The usual way of handling one part of this problem is to have receipt books that automatically make duplicates that everyone can turn in with the money he or she has raised.

If you are going to start something that involves several other people, then you might think about setting up a group to help you oversee the project, review progress, and join in publishing reports on your successes. That way the responsibility for making sure the campaign runs right will be shared, and you will have help if anything appears to be going off the rails.

If your campaign is going to go on for a while and involve a group of other people, but you do not want to set up a new nonprofit organization, then you might consider approaching an established nonprofit to act as your "fiscal sponsor" or "fiscal agent." Fiscal sponsorship means that the existing organization adopts your work as a "program" under its wing. The sponsor is ultimately responsible for the program, so you would have to follow its rules; usually the sponsor takes a small administrative fee. Donors to your project would be assured a tax deduction, though, if all the requirements of the IRS were met. In addition, you would not have to worry about administrative details like tax forms, bank deposit rules, and accounting packages.

Source: Barber, Putnam. 2005. Idealist.org. http://www.idealist.org/if/i/en/faq/404-142/110-77.

ASKING FOR STUFF

Oftentimes, individuals, businesses, and other organizations would like to give to an organization or project, but they do not have the finances or ability to do so. However, that does not mean that they have nothing to give. For this reason, fundraising campaigns need to be able to address more than cash gifts, as there are many ways to give. By knowing what resources, supplies, and services are needed in the planning and operation of the project, in-kind and volunteer donations become possible. Examples of items and services commonly donated include:

- Computers
- Printers
- Office supplies
- Books
- Medical equipment

- Office furniture
- Transportation
- Office space
- Uniforms
- Printing
- Advertising

This is but a tiny sample of the thousands of things that could possibly be needed for a public education project. If these items are listed in a proposal or in any literature developed for planning or fundraising purposes, fundraisers can discuss the need with any donor that offers assistance other than cash. However, what is collected need not be limited to the needs of the project, of course. Donated items can be sold for charity or auctioned off. Many local businesses will donate items from their inventory, which they can claim for a tax deduction, and which organizations may be allowed to sell at a significant discount to raise funding for their project. Donated advertising has been mentioned as well. The donation itself will not bring the organization closer to its fundraising goal, but it may bring it the publicity it needs to do so.

Captain Robert Moore of the Suffolk County Police Department writes,

> When solicited for money and before analyzing the merits of a proposal, potential benefactors make a number of decisions. Questions that instantly run through their minds include:
>
> - Can I afford the amount requested?
> - Do I want to give away that amount?
> - Is the amount requested in line with the scope of the work?
>
> The primary focus of each of these questions is money. The purpose, goal, or task, if considered at all, becomes a secondary issue. And a "no" response to just one of these questions usually ends the discussion. So stop thinking about money. Instead, think about what kinds of "stuff" your organization needs. (1998)

Whether you think about acquiring materials from donors as a separate campaign from the fundraising, you do need to be knowledgeable about what you need before the question is ever asked of you. As for the equipment you need for your project — if it is industry specific, you may want to ask businesses who sell or manufacture that equipment for *either* a cash or an equipment donation. Leave the option up to them. Other equipment is much more common, like desks, chairs, phones, and fax machines, for example, and almost any business is likely to have some of these items lying around that they do not need. You can even place an advertisement in the newspaper asking members of the community and businesses if they would be willing to support your organization by donating anything from the list you provide in the advertisement. Again, remember that office or other space is something that organizations often have donated to them. If your project involves training, you could have classroom space donated. If you need an operations center, an office may be given, complete with paid utilities. You can even agree on certain times or days when you will use them if that is all that you will need. Food and drinks can be donated for events. You will never know what people may give if you do not ask.

Captain Moore offers additional valuable advice in this area (see Sidebar 6.8).

Sidebar 6.8: Captain Moore's In-Kind Solicitation Advice

Have you ever driven past a home, seen items sitting at the curb awaiting garbage pickup and said to yourself, "Look at all that great stuff!" The same principle can be applied to corporations. Unless you tell them, they won't know that the furniture or equipment they consider obsolete suits your needs just fine. My station acquired desks, chairs, file cabinets, electronic typewriters, and copying machines, at no cost, from a local savings bank. The bank, like most corporations, periodically redecorates its offices and updates machinery. Because I've made them aware of our needs, a bank official always notifies us when they have something available.

A number of organizations, by mandate or corporate charter, or because of good corporate citizenship, distribute informational literature on topics such as driving and pedestrian safety, frauds and schemes, burglary and robbery prevention, preparations for natural disasters and so on. The brochures are high quality, informative, and cover virtually every topic imaginable and are often distributed to other organizations free of charge.

If you need meeting space but the rental fee isn't in your budget, call your local library or community college. Both institutions routinely make their facilities available to community groups for training sessions and seminars, and even assist with advertising and registration. They provide audio visual aids, have adequate parking and bathrooms, and are equipped with just about everything needed to host a successful event. Usually, these benefits are provided at no charge.

Other untapped resources are high schools and colleges. An increasing number of learning institutions are making community service a requirement for graduation. Schools often scramble to find opportunities for young people. It's an excellent way for your organization to help students gain experience while you gain extra personnel. I've heard of many novel programs where community members offer a service in exchange for some action or activity on the part of others. One of my favorites occurred in Contra Costa County, California, where a group of psychologists offered three hours of free therapy to anyone who turned in a handgun. There are no limits to ingenuity.

Major organizations conduct mandatory training. Fire departments, utility companies, police departments, airports, hospitals, and the Army Reserve all conduct large scale, multi-agency drills. There is no finer training available in crisis management, communications, prioritization, systems and procedures, or teamwork than drills. Let these organizations know you want to participate or observe.

A few final suggestions include the following:

- Focus your thinking on what you need or what you intend to accomplish.
- Be specific. Even when stuff is not immediately available, potential benefactors will remember you when it is.
- Benefactors will take your calls and even call you when they know money is not likely to come up in the conversation.

Source: Moore, Robert. 1998. Asking for Stuff. Community Links. Vol. 3. Winter. http://web.archive.org/web/20060103061515/http://www.communitypolicing.org/publications/comlinks/cl_4/c4_moor.htm.

A number of nonprofit organizations and associations maintain online and printed resources to help with fundraising. The following is a list of several of these organizations:

- The Foundation Center www.fdncenter.org
- Fundraiser.org http://www.fundraiser.org/
- Internet Non-profit Center http://www.nonprofits.org/
- American Fund Raising Institute http://www.amfundraising.com/
- Fundraisingweb.org http://www.fundraisingweb.org/
- Minnesota Council on Foundations http://www.mcf.org/

- Raise Funds http://www.raise-funds.com/
- Chronicle of Philanthropy http://www.philanthropy.com/
- Grass Roots Fundraising Online http://www.grassrootsfundraising.org
- Fundraising.Com www.fund-raising.com
- Contributions Magazine www.contributionsmagazine.com
- American Association of Fundraising Counsel www.aafrc.org

CONCLUSION

All public education campaigns need funding to survive. When organizations find themselves without the resources to complete, continue, or even to begin their public education efforts, what lies ahead can be daunting. Fortunately, there are countless sources of funding to which appeals may be made, as described in this chapter. The key is that the campaign that is planned and conducted always matches the funding levels that are attainable.

REFERENCES

Barber, Putnam. 2005. Idealist.org. http://www.idealist.org/if/i/en/faq/404-142/110-77.

Better Business Bureau. 2001. Donor Expectations Survey. Wise Giving Alliance. us.bbb.org.

Hall, Mary. 1988. *Getting Funded: A Complete Guide to Proposal Writing*, 3rd ed. Portland, OR: Continuing Education Publications.

Independent Sector Website, www.independentsector.org, accessed December 2008.

Moore, Captain Robert. 1998. Asking for Stuff. Community Links. Vol. 3. Winter. http://web.archive.org/web/20060103061515/http://www.communitypolicing.org/publications/comlinks/cl_4/c4_moor.htm

Poderis, Tony. 1997. 12 Things You Should Know About Setting A Capital Campaign Goal. Fund Raising Forum. http://www.raise-funds.com/998forum.html

7

Emergency Management Public Education Case Studies

CASE: MULTI-CULTURAL DISASTER PREPAREDNESS CAMPAIGN

Program Facts

Who: American Red Cross of Greater Los Angeles
What: Media Campaign "Just Be Ready: Prepare Together!"
Where: Los Angeles, California
Partners: Los Angeles County Department of Health Services
Mission: "To prevent and mitigate the public health consequences of natural or intentional emergencies for Los Angeles County residents through threat assessment, planning, improved operational readiness, and timely response."

In the summer of 2005, just prior to Hurricane Katrina's devastating strike across the U.S. Gulf Coast, the American Red Cross of Greater Los Angeles and the Los Angeles County Department of Health Services announced the launch of a yearlong disaster preparedness public service campaign called "Just Be Ready: Prepare Together!"

The campaign initially featured actor and comedian Paul Rodriguez in PSAs that aired on local radio and television stations, as well as Spanish-language radio stations.

In recognition of Los Angeles's multi-cultural, multi-lingual population, campaign partners also developed disaster preparedness public service announcements in the most common languages used by residents. These included Spanish, Chinese, Korean, Thai, Khmer, Tagalog, Vietnamese, Armenian, Russian, Arabic, and Farsi.

The campaign featured various print materials that were produced in all 12 languages, and included:

- "10 Essential Items"
- "A Children's Primer on Emergency Preparedness"
- "Your Guide to Public Health Emergency"
- "Family Emergency Supplies Kit"
- "Family Emergency Plan"
- "Emergency Plan for People with Disabilities"
- "Emergency Info for People with Special Needs"

A video titled "Make an Emergency Kit" was produced in English, Spanish, Korean, and Armenian.

Preparedness materials developed for this campaign, in all languages listed above, can be downloaded from the Los Angeles County Website at www.labt.org or publichealth.lacounty.gov/eprp/media/viewads.htm. Disaster readiness kits and other preparedness items will be purchased online from the Red Cross store at http://www.redcrossstore.org/shopper/ProdList.aspx?LocationId=1

Contact Information

Website: http://publichealth.lacounty.gov/eprp/media/index.htm
Phone: 866-999-5228

CASE: MULTI-CULTURAL DISASTER PREPAREDNESS CAMPAIGN

Program Facts

Who: American Red Cross Bay Area
What: "Prepare Bay Area" (PBA)
Where: San Francisco, California

Partners: Pacific Gas and Electric Company, businesses, faith communities, government agencies, nonprofits, schools, civic groups, and ethnic communities
Mission: "To get one million Bay Area residents committed to and demonstrating preparedness for any emergency."

In response to summary data from a new Bay Area preparedness survey conducted by the Red Cross that reported only 6% of Bay Area residents had made a disaster plan, built a kit, and received training, the American Red Cross Bay Area launched an unprecedented three-year disaster preparedness campaign entitled "Prepare Bay Area."

The program included both online and offline guidance for creating a disaster plan, making an emergency kit, and first aid and CPR training. The program set out to reach all residents of San Francisco, San Mateo, Marin, Almeda, Contra Costa, and Solano counties, but nonprofits were enlisted specifically to reach vulnerable communities of focus, including the elderly, children, persons with disabilities, the homeless, people for whom English is a second language, and low-income communities.

The campaign spread its Make a Plan, Build a Kit, and Get Trained initiative through three channels: (1) online at www.preparebayarea.org; (2) via telephone at 1-877-PREPBAY; and (3) in person through thousands of free, 30-minute specialized trainings at schools and workplaces. Paid, certified first aid and CPR trainings were also available upon request. Materials were all produced in multiple languages.

Contact Information

Website: www.redcrossbayarea.org
Phone: 1-888-4-HELP-BAY

CASE: PREPARING CHILDREN FOR EMERGENCIES

Program Facts

Who: Life EMS Ambulance
What: One-day summer camp for children
Where: Kalamazoo, Michigan

Partners: The Safe Kids Coalition, Prevention Works, and the American Heart Association

Mission: To give children "an increased awareness of safety and prevention."

In the summer of 2008, Life EMS ambulance in Kalamazoo, Michigan, offered four separate sessions of Camp 9-1-1, a one-day summer camp that aims to teach children what to do to get immediate help in emergency situations and other ways to keep themselves safe.

Participants did not receive CPR certification, but they did get basic instructions in CPR and first aid and they learned what rescuers and first responders do in emergencies. The program also included segments on bike safety training and drug and alcohol prevention.

Students were able to try out the equipment that firefighters and emergency workers use, and practice CPR moves on a dummy. They learned how to call 911 and how to get low to the ground and escape from a burning trailer. Students also received healthy snacks, brochures, and DNA kits that included fingerprinting and hair sample kits provided by the American Heart Association. To register or for more information, call Kimberly Caton at (269) 373-3104. For matters pertaining to camps at the museum, call Elizabeth Barker at (269) 373-7965.

CASE: ORGANIZED TRAINING FOR COMMUNITIES

Program Facts

Who: Los Angeles Fire Department (LAFD)

What: Community Emergency Response Team (CERT)

Where: Los Angeles, California

Partners: LAFD, FEMA, and Emergency Management Institute

Mission: To prepare the public to be able to rely on each other in order to meet their immediate lifesaving and life-sustaining needs in the case of an emergency situation in which first responders are unable to meet the demand for these services.

In 1987, the Whittier Narrows earthquake emphasized the threat of a major disaster in California, and confirmed the need to train citizens to meet their own immediate needs. As a result, the LAFD created a Disaster Preparedness Division to train citizens.

The program was highly successful, and, in time, FEMA decided that it should be made available to all communities throughout the nation. In 1994, the Emergency Management Institute and the LAFD expanded CERT materials to make them applicable to all hazards.

Disasters vary greatly from location to location, so training varies based on the probabilities of different types of disasters. Members are taught to assess their own needs and the needs of the immediate environment first (e.g., family, neighbors, and community). The certification process for CERT requires 17 hours of training dispensed over a seven-week period and includes disaster preparedness, light urban search and rescue, fire safety, basic medical principles, teamwork and the incident command system, disaster psychology, and terrorism preparedness, and there is a final exercise.

Today, there are more than 2300 CERTs registered nationwide. They are sometimes called in to designated staging areas to fill different assigned roles based on needs. Instructions on how to start a CERT program, how to apply for CERT, a state directory and a map of nearby CERTs, frequently asked questions, all training materials, video materials, supplemental information, and contact information can be found on the organization's Website: http://www.citizencorps.gov/cert/about.shtm.

Contact Information

Website: http://www.citizencorps.gov/cert/about.shtm
Email: cert@dhs.gov

CASE: CHILDREN'S "EDUTAINMENT" PROGRAM

Program Facts

Who: First responders in various cities throughout the United States
What: Children's public safety program: Fire Pals
Where: Elementary schools in various cities throughout the United States
Partners: Firefighters, paramedics, police officers, sheriff's deputies, and other public safety personnel; the CDC is mentioned as the funding source on the Webpage of the Fire PALS program for Waterloo, Iowa.

Mission: "The goal of our Fire PALS Program is to teach school children, adults, elderly, and those in our community with special needs the skills necessary to protect themselves and their families by responding promptly and effectively when confronted with a fire or life safety hazard."

Public safety personnel go to schools to teach public safety messages through music, skits, and puppets, including "Stop drop and roll," "Always wear a helmet," "Have an exit plan and two ways out (of a burning house)," and "Get out, stay out, and have a safe meeting place outside during a fire."

The Waterloo, Iowa, program has a Webpage that includes information for children and adults, contact information, safety tips, program goals, course descriptions, and downloads: http://www.firepals.com/adults.php. A newspaper article in the *Oroville Mercury Register* about one of these programs in California, the CAL Fire-Butte County's Public Information Officer, noted that there are documented cases of lives being saved because children remembered Fire Pals' life saving messages.

Contact Information

Waterloo Fire PALs Website: http://www.firepals.com/adults.php

CASE: DISASTER PREPAREDNESS AT RELIGIOUS INSTITUTIONS

Program Facts

Who: Brooklyn city officials and New York Disaster Interfaith Services
What: Emergency information Website: www.howcalm.org
Where: Brooklyn, New York
Partners: Brooklyn city officials, faith-based organizations, the American Red Cross, and FEMA
Mission: To provide information about disaster preparedness to religious institutions that can then relay those details to congregants.

Faith-based organizations possess a unique opportunity to reach certain portions of the population that may not be reached through traditional emergency preparedness promotion techniques. New York Disaster

Interfaith Services recognized the importance of houses of worship as critical centers of community life in Brooklyn, New York, and encouraged the creation of a Website devoted to providing religious institutions, schools, and service providers with information about how to aid congregants in times of crisis and offer assistance during rebuilding efforts after a disaster. Additionally, the Website offers details about evacuation zones and gathers information about which houses of worship maintain facilities that could accommodate evacuees if there is a full-scale emergency.

HOWCALM stands for House of Worship Communitywide Asset & Logistics Management. The HOWCALM site is free, and organizations specializing in preparing communities are welcomed. Currently, the system is only offered to residents of New York City, but the Website says that it will launch in other locations outside New York City at a future date.

CASE: EARTHQUAKE READINESS TAUGHT TO A POPULATION THAT SPEAKS ENGLISH AS A SECOND LANGUAGE

Program Facts

Who: CERT
What: Earthquake readiness program taught in Mandarin
Where: San Francisco, California
Partners: Sponsored by the city of San Francisco, local Chinese-American residents, Tzu Chi Foundation
Mission: To teach the emergency preparedness skills promoted by the CERT program to Mountain View's sizeable segment of the population whose primary language is Mandarin.

In an effort to promote emergency preparedness for all members of the community, Mountain View in California developed CERT training programs in additional languages so that more members of the community could receive this vital information in their native language. The program was developed first by identifying Chinese-American residents who had taken the CERT training in English. They then borrowed a training manual from San Francisco and trained the Chinese-American volunteers to be trainers themselves. Volunteers and the Tzu Chi Foundation helped them to translate materials from the manual as well as some of their own materials.

In class, residents learned about team building during an emergency, went over the basics of emergency response, learned how to inspect an injured person from head to toe, and learned how to inspect a fire.

Lynn Brown, Mountain View's coordinator of emergency services, told a reporter from the *San Jose Mercury News* at one of the sessions, "Most everyone here speaks some English. But it's easier for them to grasp the concepts quicker. It makes a difference."

The information taught in this program is the same as the information provided in the English version of CERT (discussed earlier).

CASE: CITYWIDE PREPAREDNESS EFFORT

Program Facts

Who: Bothell Fire and EMS Services

What: Operation: Every House Prepared

Where: Bothell, Washington, and Snohomish County Fire Protection District 10

Partners: The distributions were funded through an Emergency Management Performance Grant awarded through the state's military department and the federal Department of Homeland Security; packets were assembled and distributed by community volunteers

Mission: To help residents of Bothell, Washington, prepare for disasters.

In March 2007, volunteers in the city of Bothell, Washington, compiled and hand-delivered 15,000 emergency preparedness packets to households in the city and the Snohomish County Fire Protection District 10. Volunteers distributing the packets wore orange identification jackets issued by the city. Households that could not be reached through door-to-door services received packets through the mail.

Each packet contained a multi-lingual sheet explaining the packet, a flier with contacts for various services, a letter from the city manager, a disaster preparedness guide produced by the military department, a vendor list to help residents obtain emergency supplies, and a volunteer list recognizing community members assisting with the effort.

A copy of the packet and information about the campaign can be obtained online at the City of Bothell Office of Emergency Preparedness Website: http://www.ci.bothell.wa.us/dept/fd/disaster/disaster.html.

Contact Information

Website: http://www.ci.bothell.wa.us/dept/fd/disaster/awareness.html
Phone: A press release about the campaign directed questions to public information officers with Bothell Fire and EMS services at 425-489-336.

CASE: VOLUNTEER EMERGENCY PREPAREDNESS PROGRAM

Program Facts

Who: Sunnyvale, California, city government and its citizens
What: SNAP (Sunnyvale Neighborhoods Actively Prepare)
Where: Sunnyvale, California
Partners: City government and citizens of Sunnyvale, California
Mission: "To work with people in neighborhoods to ensure that they were better able to cope with the aftermath of earthquakes."

SNAP is a volunteer program developed on a neighborhood-by-neighborhood basis to improve the abilities of people in the community to cope with the aftermath of earthquakes. Each self-defined neighborhood composed of 35 to 50 homes is represented by a captain and six disaster task committees. Captains coordinate action-based committees that prepare to use citizens as resources in times of need. Each committee represents one of the key concerns in the event of a disaster:

- Communications
- Damage assessment
- First aid
- Safety and security
- Search and rescue
- Sheltering and special needs

Some captains are also part of a citywide SNAP steering committee that works directly with city staff.

The program offers five different workshops and a full set of informational materials:

- Basic committee training
- Topic training
- Captain training

- First aid/CPR training
- Advanced urban search and rescue training

Training provides education and information, and committees provide a joint response of citizens and government. Citizens become part of the decision-making process as well as participate in community improvement.

The program is funded by tax dollars. FEMA is using SNAP as a model for residential emergency preparedness.

Contact Information

Website: http://sunnyvale.ca.gov/Departments/Public%2BSafety/Office%2Bof%2BEmergency%2BServices/SNAP%2BPS.htm
Phone: (408) 730-7117

CASE: TEACHING EMERGENCY PREPAREDNESS IN SCHOOLS

Program Facts

Who: New England Schools
What: The Student Tools for Emergency Planning (STEP) program
Where: New England
Partners: FEMA and state emergency management agencies with local schools in New England
Mission: To teach basic emergency preparedness skills.

In 2008, FEMA and state emergency preparedness agencies teamed up with New England schools to launch a pilot program for New England fourth-grade students to teach basic emergency preparedness skills. The STEP program provides ready-to-teach preparedness lessons for teachers to empower students to encourage their families to make home emergency kits and communication plans.

In addition to the lesson plan, some schools also were provided with items for students to make their own "starter kits," including a water bottle, snack bar, emergency whistle, Mylar blanket, and carrying bag.

The family plan promoted by the program encouraged students to agree with their families on an alternate meeting place outside of the home, designate a person to call in case of an emergency (often in a different state), and for each family member to carry around cards with those important facts as well as work and school numbers.

More information about the emergency preparedness kits and communications plans can be found at www.fema.gov/plan/prepare/supplykit.shtm and www.fema.gov/plan/prepare/commplan.shtm

CASE: EMERGENCY PREPAREDNESS IN PUBLIC TRANSPORTATION

Program Facts

Who: Washington Metropolitan Area Transit Authority
What: Emergency evacuation maps help riders find alternate bus or rail services
Where: Washington, D.C.
Partners: Washington Metrorail stations
Mission: "An ongoing initiative to improve communications with customers, particularly during emergencies."

Washington Metro created site-specific emergency evacuation maps that customers can take with them if they have to evacuate a rail station or choose to find an alternate route during a lengthy service disruption. Maps were customized for each of Metrorail's 117 station mezzanines. They list all bus services available near the station and provide walking directions to the closest Metrorail station, along with the distance and estimated walking time. A "You Are Here" icon is provided to assist passengers in situating themselves on the map.

Each mezzanine station initially received 5000 copies of the individualized 8.5 × 11-inch paper maps. Metro staff replenishes the maps periodically or as inventory gets low. The maps were created as part of a larger initiative to improve maps in the Metrorail system in general. Maps are also provided on the Washington Metropolitan Transit Authority Website.

Contact Information

Website: http://www.wmata.com/rail/maps/map.cfm

CASE: FEMA PREPARES CHILDREN FOR DISASTERS

Program Facts

Who: FEMA for Kids
What: Educational program for kids
Where: Westwago, Louisiana
Partners: FEMA
Mission: To help children who were hit hardest by Hurricane Katrina to prepare for possible future disasters.

In 2007, FEMA distributed activity books to public and private schoolchildren in Orleans, Jefferson, and St. Martin parishes, all of which were affected by Hurricane Katrina. One school that was hit particularly hard by the disaster, Westwago Elementary, used the program to help students get prepared if they have to face another disaster.

The books are primarily targeted at elementary school children, and they feature tasks such as how to make a family disaster plan, what to pack in a disaster kit, what to do in case of a fire, and more.

Materials can be viewed at the FEMA for Kids Website: www.fema.gov/kids. Free books, posters, videos, and other educational materials are available to anyone who calls 1-800-480-2520 or goes to the Website.

Contact Information

Website: www.fema.gov/kids
Phone: 1-800-480-2529

CASE: DISASTER PREPAREDNESS IN SCHOOLS

Program Facts

Who: Public Affairs staff from Mississippi Transitional Recovery Office
What: Elementary School Program: FEMA in Schools
Where: Mississippi Gulf Coast Schools
Partners: Schools
Mission: To educate elementary school-aged children to protect themselves, their families, and their pets during an emergency.

FEMA's public affairs staff created a school program with the aim to increase knowledge in communities, starting with children, about how to protect themselves, their families, and their pets during an emergency. The program teaches disaster preparedness through various hands-on activities. Such activities include:

- An exercise in which children create their own virtual disaster preparedness kits and explain the significance in each item in the kit
- A trip to the library to research disaster preparedness
- A puppet show for kindergarten through second grade
- Mini-classes, such as pet preparedness

All students who participate in the program receive a bag with reading materials in it for themselves and their parents. Much of the information provided by the program can be found at the FEMA Webpage for kids: www.fema.gov/kids. The Website, established in 1997, features colorful graphics, games, quizzes, and original stories to help children absorb as much information as possible.

Contact Information

FEMA for Kids Website: www.fema.gov/kids
Phone: 228-594-3068 (Cheryl Bozeman or Neily Chapman)

CASE: EMERGENCY PREPAREDNESS IN NEIGHBORHOODS

Program Facts

Who: Tiburon Peninsula Disaster Preparedness Taskforce
What: Local preparedness effort: The Get Ready Program
Where: Marin County, California
Partners: Funded by a grant from the State Department of Homeland Security and the Board of Supervisors
Mission: To teach residents how to prepare for, survive through, and recover from a disaster.

In 2007, the Tiburon Peninsula Disaster Preparedness Taskforce launched the Get Ready Program, a disaster survival program for residents of Marin County, California. Bright yellow signs affixed to schools, fire stations, and roadside fences were posted from Stinson Beach to San Rafael, encouraging people to "Get Ready" for the next earthquake, flood, fire, landslide, or other natural disaster.

The campaign's objective was to promote a free, two-hour-long program founded in southern Marin featuring everything one needs to know about saving oneself in a disaster.

Promotions for the class included banners, signs, posters, Golden Gate Transit bus advertisements, cocktail napkins, cable television public service announcements, and an airplane with a trailing banner.

Contact Information

Website: www.getreadymarin.org
Telephone: 485-3131

CASE: PUBLIC PRIVATE PARTNERSHIP FOR DISASTER PREPAREDNESS

Program Facts

Who: Tulsa Citizen Corps Partners and McDonald's
What: Mayor's Citizen Corps Month
Where: 32 McDonald's restaurants in Tulsa, Broken Arrow, Bixby, Catoosa, Glenpool, Jenks, Owasso, and Sand Springs, Oklahoma
Partners: Tulsa Citizen Corps Partners and the McDonald's Corporation
Mission: "To provide our customers with the tools they need to 'get a plan, get a kit, and get involved,' to prepare themselves and their families for disaster emergencies."

In June 2003, Tulsa mayor Bill LaFortune, president of McDonald's Tulsa Co-op of Restaurant Owners Dwayne Sampson, and Ronald McDonald held a press conference at a local McDonald's restaurant to kick off "Mayor's Citizen Corps Month."

The four Mayor's Citizen Corps Month initiatives designed to bring family preparedness information to McDonald's customers and the community included:

1. Family preparedness displays at 32 McDonald's restaurants in Tulsa, Broken Arrow, Bixby, Catoosa, Glenpool, Jenks, Owasso,

and Sand Springs to remain in the restaurants for the month. They included information on floods, tornadoes, safe rooms, disaster preparedness, and Citizen Corps. The displays were tailored to the individual neighborhood for each restaurant location, with special Neighborhood Safety Assessment maps created at INCOG (Indian Nations Council of Governments).

2. Citizen Corps tray liners featuring the message "Get a plan! Get a kit! Get involved!" being used at 64 restaurants in the northeast Oklahoma demographic area.
3. A special family preparedness brochure including the "Get a plan! Get a kit! Get involved!" theme was placed in the same 64 restaurants.
4. McDonald's also joined Tulsa Partners sponsoring the Flag Day Pancake Breakfast, Tulsa Partners' first fundraising project. McDonald's offered a $600 cash donation, condiments and paper products, and 500 backpacks, plus items such as coloring books and crayons, to give away to children at the event. Ronald McDonald, Grimace, and Hamburglar attended the breakfast, and Ronald presented his "Family Safety Show" to young audiences.

Contact Information

A press release about the event can be found at http://www.citizencorps.gov/news/press/2003/2003_05_mcd.shtm.
Citizen Corps Website: http://www.citizencorps.gov/index.shtm.

CASE: NATIONAL PUBLIC EDUCATION EFFORT

Program Facts

Who: Civil Defense Minister Rick Barker
What: Nationwide preparedness campaign: "Get Ready Get Thru"
Where: New Zealand
Partners: Television and radio stations, the consortium (Clemengers, Senate, and Colmar Brunton), schools, seeking others
Mission: "To boost public awareness and understanding of the need to prepare to face disasters by having a plan, and taking steps to be better prepared."

In June 2006, New Zealand Civil Defense Minister Rick Barker launched a nationwide campaign that urged New Zealanders to "Get Ready Get Thru." The mass media campaign included radio, television, and print ads to encourage New Zealanders to prepare for disasters. The key messages to this campaign were listed on New Zealand's Department of Internal Affairs Webpage as:

- In a disaster, essential services will be disrupted. Emergency services and civil defense staff will be doing their job but help cannot get to everyone as quickly as they may need it.
- Each and every one of us needs to take responsibility to plan to look after those dependent on us.
- We need to take steps now to be prepared to look after ourselves for up to three days or more. (http://www.dia.govt.nz/diawebsite.nsf)

In addition to the mass media portion of the initiative, other public initiatives included:

- A school program entitled "What's the Plan Stan?" was sent out to more than 3000 primary and intermediate schools. Fifteen professional development workshops were held around the country and were attended by more than 500 teachers and principals, and civil defense staff.
- A Website, www.getthru.govt.nz, was created to provide user-friendly information and advice for the public on what they should do to be prepared. The Website also offered links to the nearest council so that people could easily access information specific to their region.
- The agency also created a Public Education Toolbox, (PE)Toolbox, to provide those tasked with developing public education programs for CDEM (Civil Defence and Emergency Management) groups with resources, such as templates, written materials, articles and media releases, communications strategies, a photo database, and examples of programs undertaken by others.
- The program included an intensive evaluation component of quantitative benchmark research to understand current national levels of awareness, understanding, and preparedness.
- The program also became very involved with the promotion of the annual Disaster Preparedness Week.

Contact Information

Website: www.getthru.govt.nz
Email: For further information on the National Public Education Program, please contact chandrika.kumaran@dia.govt.nz
To obtain additional copies of "What's the Plan Stan?" materials, contact chris.baylis@dia.govt.nz.

APPENDIX: WEBSITES AND DOWNLOADABLE GUIDES FOUND ON THE INTERNET

Public Education/Communicating with the Public

CDCynergy: Emergency Risk Communication
http://www.orau.gov/cdcynergy/
Website provides a framework for public health communicators to plan for, respond to, and evaluate communication efforts during a terrorist event. Users are able to download a version of the Centers for Disease Control and Prevention's *Crisis and Emergency Risk Communication: Be First, Be Right, Be Credible.* Both the book and the Website detail the application of emergency risk communication principles.

Improvement and Development Agency
http://www.idea.gov.uk/idk/core/page.do?pageId=7816073
Communications toolkit linked to the Webpage of Improvement and Development Agency for local government (I&DEA), a UK company that works for local government improvement so councils can serve people and places better. The toolkit includes links to articles about reputation, branding and place shaping, corporate communication strategy, managing media relations, community engagement, legal issues, and case studies.

U.S. Army Center for Health Promotion and Preventive Medicine
http://chppm-www.apgea.army.mil/documents/TG/TECHGUID/TG270.pdf
Technical guide created by an international working group hosted by the U.S. Army Center for Health Promotion and Preventive Medicine to provide guidance on effectively educating the public about the health hazards of excessive exposure to ultra-violet radiation (UVR). An international group

of experts working under the auspices of the World Health Organization (WHO) INTERSUN program contributed to the guide. Segments include: Establishing an Educational Program on UVR Hazards and Protection, Creating a Supportive Environment for UVR Education, and Evaluating the Success of the UVR Educational Program. Examples of all materials are also provided.

U.S. Department of Health and Human Services
http://www.hhs.gov/disasters/press/newsroom/mediaguide/
Field guide posted by the U.S. Department of Health and Human Services designed to serve as a guide for media to quickly and clearly communicate terrorism and public health emergency messages to the public. Field guide sections include: Introduction; Biological Agents; Chemical Agents; Radiation Emergencies; Defining Public Health Roles, Programs, and Terms; Web Sites; and Contact Lists, Checklists, and Additional Tools.

U.S. Department of Health and Human Services
http://www.hhs.gov/disasters/press/newsroom/leadersguide/index.html
A guide on public health response to emergencies that is specifically tailored for public officials (e.g., mayors, governors, county executives, emergency managers) and first responders (police, fire, EMS). Chapters include: Introduction, Public Health Response, The Key Functions of Federal Government Public Health Agencies in an Emergency, Food Security and Food Supply, Environmental Safety and Testing, Leading Through Communication: The Role of Risk Communication During a Terrorist Attack or Other Public Health Emergency, Legal and Policy Considerations, Taking Care of Yourself and Each Other, Conducting Exercises for Preparedness, Post-Event: Leading Your Community Towards Recovery and Resiliency, and Bibliography.

Social Marketing

Department of Health and Human Services: Centers for Disease Control and Prevention
http://www.cdc.gov/healthmarketing/
Health marketing Webpage created by the National Center for Health Marketing. Links include What Is Health Marketing?, Health Marketing Basics, Health Marketing in Action, Research and Evaluation, Resources

and Tools, eHealth Marketing, Global Health Marketing, Professional Development, Partnerships, and About CDC's National Center for Health Marketing.

Institute for Social Marketing
http://www.ism.stir.ac.uk/index.htm
Website for the Institute of Social Marketing, an establishment with 28 years of experience with the study and dissemination of social marketing theory and practice and eventual creator of the National Social Marketing Centre in England. Links include information on what is social marketing, case studies of social marketing in practice, project descriptions, and links to other social marketing Websites.

National Social Marketing Centre
http://www.nsms.org.uk/public/default.aspx
Website of a strategic partnership between the Department of Health in England and Consumer Focus (formerly the National Consumer Council). In 2006, the latter launched *It's Our Health,* the first-ever national review of health-related campaigns and social marketing in England. Links from this page include: What Is Social Marketing?; Why Is Social Marketing Important?; Resources, Documents, and Presentations; and Health Literacy and Social Marketing.

Social Marketing: A Resource Guide
http://www.turningpointprogram.org/Pages/pdfs/social_market/social_marketing_101.pdf
A basic guide to using social marketing put out by Turning Point, a former network of 23 public health partners started by the Robert Wood Johnson Foundation and the W. K. Kellogg Foundation to transform and strengthen the public health system in the United States by making it more community based and collaborative. The information in the guide was put out to benefit public health program planners, public information and public affairs specialists, health educators, health communicators, and health and wellness promoters.

Weinreich Communications Social Marketing Articles
http://www.social-marketing.com/library.html
A collection of social marketing Websites compiled by Weinreich Communications, a company founded by Nedra Kline Weinreich, author of the often-cited and widely used *Hands-On Social Marketing: A Step-by-Step*

Guide. Links include articles on "What is Social Marketing?", "Building Social Marketing Into Your Program," "Getting Your Message Out Through the Media," "Research in the Social Marketing Process," "Integrating Qualitative and Quantitative in Social Marketing Research," and "Social Marketers in the Driver's Seat: Motorsport Sponsorship as a Vehicle for Tobacco Prevention, and Strategic Social Marketing for Non-Profits."

INDEX

F

K

L

M

N

T

U

V

W

Y